Black Boxes

Black Boxes

How Science Turns Ignorance into Knowledge

MARCO J. NATHAN

OXFORD
UNIVERSITY PRESS

OXFORD
UNIVERSITY PRESS

Oxford University Press is a department of the University of Oxford. It furthers
the University's objective of excellence in research, scholarship, and education
by publishing worldwide. Oxford is a registered trade mark of Oxford University
Press in the UK and certain other countries.

Published in the United States of America by Oxford University Press
198 Madison Avenue, New York, NY 10016, United States of America.

Library of Congress Cataloging-in-Publication Data
Names: Nathan, Marco J., author.
Title: Black boxes : how science turns ignorance into knowledge / Marco J. Nathan.
Description: New York, NY : Oxford University Press, [2021] |
Includes bibliographical references and index.
Identifiers: LCCN 2021001998 (print) | LCCN 2021001999 (ebook) |
ISBN 9780190095482 (hardback) | ISBN 9780190095505 (epub)
Subjects: LCSH: Science—Philosophy. | Science—Methodology. |
Ignorance (Theory of knowledge)
Classification: LCC Q175 .N38 2021 (print) | LCC Q175 (ebook) | DDC 501—dc23
LC record available at https://lccn.loc.gov/2021001998
LC ebook record available at https://lccn.loc.gov/2021001999

DOI: 10.1093/oso/9780190095482.001.0001

1 3 5 7 9 8 6 4 2

Printed by Integrated Books International, United States of America

Dedicated, with love, to Jacob Aaron Lee Nathan.
Benvenuto al mondo.

Like all general statements, things are not as simple as I have written them, but I am seeking to state a principle and refrain from listing exceptions.

—Ernest Hemingway, *Death in the Afternoon*

Contents

Preface

Every intellectual project has its "*eureka!*" moment. For this book, it happened a few summers ago during a lonely walk on the beach of Marina di Campo, on the beautiful island of Elba, off the Tuscan coast. I suddenly started seeing a guiding thread, a *leitmotiv*, that connected much of my reflections on the nature of science since I started working on these issues back in graduate school. Simply put, it dawned upon me how I viewed most scientific constructs as *placeholders*. Explanations, causal ascriptions, dispositions, counterfactuals, emergents, and much else. They could all be viewed as boxes, more or less opaque, standing in for more detailed descriptions. That got me thinking about how to provide a more unified account that puts all the tiles of the mosaic together. At the same time, I had realized that the very concept of a black box, so frequently cited both in specialized and popular literature, has been unduly neglected in philosophy and in the sciences alike. This book is the result of my attempts to bring both insights together, in a more or less systematic fashion.

The intellectual journey sparked by my preliminary reckoning on a sandy beach has taken several years to complete. Along the way, I have been honored by the help and support of many friends and colleagues. Philip Kitcher and Achille Varzi encouraged me to pursue this project from the get-go. Many others provided constructive comments on various versions of the manuscript. I am especially grateful to John Bickle, Giovanni Boniolo, Andrea Borghini, Stefano Calboli, Guillermo Del Pinal, George DeMartino, Enzo Fano, Tracy Mott, Emanuele Ratti, Sasha Reschechtko, Michael Strevens, and Anubav Vasudevan for their insightful comments. A special thank you goes to Mika Smith, Roscoe Hill, Mallory Hrehor, Naomi Reshotko, and, especially, Bill Anderson, all of whom struggled with me through several drafts and minor tweaks, in my endless—futile, but no less noble—quest for clarity and perspicuity.

Over the years, the University of Denver and, especially, the Department of Philosophy have constantly provided a friendly, supporting, and stimulating environment. Various drafts of the manuscript were presented as part of my advanced seminar *Ignorance and Knowledge in Contemporary Scientific*

Practice. I am indebted to all the students who shared thoughts, comments, and frustrations with me—on this note, a shoutout to Blake Harris, Olivia Noakes, and Jack Thomas. Two reading groups, one at the University of Milan and one at the University of Denver, have been quite helpful. Bits and pieces of this project have been presented, in various forms, at several institutions, too many to adequately acknowledge. Audiences across the world have provided invaluable feedback.

I am very grateful to the production staff at Oxford University Press—especially Peter Ohlin and his team for believing in the project from the very beginning and their constant support, as well as Dorothy Bauhoff's careful proofreading. Two anonymous reviewers provided precious feedback. Also, a heartfelt "Grazie!" to Stefano Mannone of *MadMinds* for crafting the images and bearing with all my nitpicky-ness.

Finally, none of this would have been possible without the unfaltering help, patience, and support of my extended family. Lots of love to my wife Heather, my parents Alida and Jacob, my sister Shirli, my brother David, my two "adopted" brothers Federico and Matteo, and my nieces and nephews: Martina, Virginia, Alexander, and Nicoló. Thanks for brightening my days. While this book was slowly seeing the light, my own life has been rocked and blessed by the birth of my son, Jacob. This work is dedicated to him: welcome to the world!

1

Bricks and Boxes

Knowledge is a big subject.
Ignorance is bigger.
And it is more interesting.

—Stuart Firestein, *Ignorance*, p. 10

§1.1. The Wall

At the outset of a booklet, aptly entitled *The Art of the Soluble*, the eminent
biologist Sir Peter Medawar characterizes scientific inquiry in these terms:

> Good scientists study the most important problems they think they can
> solve. It is, after all, their professional business to solve problems, not
> merely to grapple with them. The spectacle of a scientist locked in combat
> with the forces of ignorance is not an inspiring one if, in the outcome, the
> scientist is routed. (Medawar 1969, p. 11)

Many readers will find this depiction captivating, intuitive, perhaps even
self-evident. What is there to dispute? Is modern science not a spectacularly
successful attempt at solving problems and securing knowledge?

Yes, it is. Still, one could ask, what makes the spectacle of a scientist
locked in combat with the forces of ignorance so uninspiring? Why is it that
we seldom celebrate ignorance in science, no matter how enthralling, and
glorify success instead, regardless of how it is achieved? To be fair, we may
excuse ignorance and failure, when they have a plausible explanation. But ignorance is rarely—arguably never—a goal in and of itself. Has a Nobel Prize
ever been awarded for something that was *not* accomplished?

The key to answering these questions, and for understanding Medawar's
aphorism, I maintain, is to be sought in the context of a long-standing
image of science that has, more or less explicitly, dominated the scene well
into the twentieth century. The goal of scientific inquiry, from this hallowed

Black Boxes. Marco J. Nathan, Oxford University Press. © Oxford University Press 2021.
DOI: 10.1093/oso/9780190095482.003.0001

perspective, is to provide an accurate and complete description of the universe, or some specific portion of it. Doing science, the story goes, is analogous to erecting a wall. Or, borrowing a slightly artsier metaphor, it is like tiling a mosaic. Walls are typically made out of stone or clay. Mosaics are crafted by skillfully assembling tiles. The bricks of science, its fundamental building blocks, are facts. Science is taken to advance through a slow, painstaking, constant discovery of new truths about the world surrounding us.

Intuitive as this may sound, we must be careful not to read too much into the simile. First, there are clear discrepancies of sheer scale. The task confronting the most creative builder or ambitious artist pales in comparison to the gargantuan endeavor that the scientific community, taken in its entirety, is striving to accomplish. Second, despite an impressive historical record, the current outlook is, at best, very partial. What we have discovered about the universe is dwarfed by what we do not yet know. Third, many of our findings are bound to be inexact, at some level of description, and, when we turn our attention to the most speculative theories, they may even be widely off the mark. In short, the likening of buildings, works of art, and science should be taken with more than just a grain of salt.

Still, the old image is suggestive, optimistically indicating that overall trends and forecasts are positive. Despite a few inevitable hiccups, our grand mosaic of the universe is slowly but surely getting bigger by the day. The goal of the scientist is to identify a problem, solve it, and discover some new facts. This, I maintain, is the backdrop to Medawar's quote. Paraphrasing Pink Floyd, all in all it's just another brick in the wall.

Here is where things start to get more interesting. Over the last few decades, a growing number of scientists, philosophers, historians, and sociologists have argued that the depiction of science as progressive accumulation of truth is, at best, a simplification. Consequently, Medawar's suggestive image of the scientist battling against the evil forces of ignorance has gradually but inexorably fallen out of style. Much has been said about the shortcomings of this old perspective, and some of these arguments will be rehearsed in the ensuing chapters. The point, for the time being, is that very few contemporary scholars, in either the sciences or the humanities, would take it at face value any longer. The wall has slowly crumbled.

What is missing from the traditional picture of science as accumulating truths and facts? Simply put, it is lopsided. Knowledge, broadly construed, is a constituent—and an important one at that—of the scientific enterprise. Yet, it is only one side of the coin. The other side involves a mixture of what we do not know, what we cannot know, and what we got wrong. In a multifaceted word, what is lacking from the old conception of science is the productive role of *ignorance*. But what does it mean for ignorance to play a "productive" role? Could ignorance, at least under some circumstances, be positive, perhaps even preferable to the corresponding state of knowledge?

The inevitable presence of ignorance in scientific practice is neither novel nor especially controversial. Generations of philosophers have extensively discussed the nature and limits of human knowledge and their implications. Nevertheless, ignorance was traditionally perceived as a hurdle to be overcome. Over the last few decades it has turned into a springboard, and a more constructive side of ignorance began to emerge. Allow me to elaborate.

In a recent booklet, Stuart Firestein (2012, p. 28), a prominent neurobiologist, remarks that "Science [. . .] produces ignorance, possibly at a faster rate than it produces knowledge." At first blush, this may sound like a pessimistic reading of Medawar, where the spectacle of a scientist routed in combat by forces of ignorance becomes uninspiring. Yet, as Firestein goes on to clarify, this is not what he has in mind: "We now have an important insight. It is that the problem of the unknowable, even the really unknowable, may not be a serious obstacle. The unknowable may itself become a fact. It can serve as a portal to deeper understanding. Most important, it certainly has not interfered with the production of ignorance and therefore of the scientific program. Rather, the very notions of incompleteness or uncertainty should be taken as the herald of science" (2012, p. 44).

Analogous insights abound in psychology, where the study of cognitive limitations has grown into a thriving research program.[1] Philosophy, too, has followed suit. Wimsatt (2007, p. 23) fuels his endeavor to "re-engineer philosophy for limited beings" with the observation that *"we can't idealize deviations and errors out of existence in our normative theories because they are central to our methodology. We are error prone and error tolerant*—errors are unavoidable in the fabric of our lives, so we are well-adapted to living with and learning from them. We learn more when things break down than when they work right. *Cognitively speaking, we metabolize mistakes!"* In short, ignorance pervades our lives.

Time to connect a few dots. We began with Medawar's insight that science is in the puzzle-solving business. When one grapples with a problem and ends up routed, it is a sign that something has gone south. All of this fits

---

[1] See, for instance, Gigerenzer (2007) and Kahneman (2011).

in well with the image of science as a cumulative, brick-by-brick endeavor, which has dominated the landscape until and for the better part of the twentieth century. Still, scholars are now well aware that ignorance is not always or necessarily a red flag. The right kind of mistakes can be a portal to success, to a deeper understanding of reality. As Wimsatt puts it, while some errors are important, others are not. Thus, "what we really need to avoid is not errors, but significant ones from which we can't recover. Even significant errors are okay as long as they're easy to find" (2007, p. 24).

All of this well-known. But what does it mean to claim that ignorance may become a portal to deeper understanding? Early on in the discussion, Firestein acknowledges that his use of the word "ignorance" is intentionally provocative. He subsequently clarifies his main point by drawing a distinction between two very different kinds of ignorance. As he elegantly puts it, "One kind of ignorance is willful stupidity; worse than simple stupidity, it is a callow indifference to facts of logic. It shows itself as a stubborn devotion to uninformed opinions, ignoring (same root) contrary ideas, opinions, or data. The ignorant are unaware, unenlightened, uninformed, and surprisingly often occupy elected offices. We can all agree that none of this is good" (2012, p. 6). Nevertheless, he continues, "there is another, less pejorative sense of ignorance that describes a particular condition of knowledge: the absence of fact, understanding, insight, or clarity about something. It is not an individual lack of information but a communal gap in knowledge. It is a case where data don't exist, or more commonly, where the existing data don't make sense, don't add up to a coherent explanation, cannot be used to make a prediction or statement about some thing or event. This is knowledgeable ignorance, perceptive ignorance, insightful ignorance. It leads us to frame better questions, the first step to getting better answers. It is the most important resource we scientists have, and using it correctly is the most important thing a scientist does" (2012, pp. 6–7).

Needless to say, it is the latter notion of ignorance—knowledgeable, perceptive, insightful ignorance—that I intend to further explore throughout this book. Recognizing that failure can be important for the advancement of science, and that not all ignorance is made equal, is a small step in a long, tortuous stride. Once the spotlight is pointed in this direction, many provocative questions arise. What is productive ignorance? What is its role in scientific research? Is it merely an indicator of where further work needs to be done, or are there really scientific questions where ignorance may actually be preferable to knowledge? Can ignorance teach us something, as opposed to merely showing us what is missing? What makes some errors fatal and others negligible or even useful? To put all of this in general terms, how does science turn ignorance and failure into knowledge and success?

Much will be said, in the chapters to follow, about the nature of productive ignorance, what distinguishes it from "a stubborn devotion to uninformed opinions," and how it is incorporated into scientific practice. Before doing so, in the remainder of this section, I want to draw attention to a related issue, concerning pedagogy: how science is taught to bright young minds. If, as Firestein notes, ignorance is so paramount in science, why is its role not explicitly incorporated into the standard curriculum?

As noted, most scholars no longer take seriously the cumulative ideal of science as the slow, painstaking accumulation of truths. Now, surely, there is heated debate and disagreement on how inaccurate this picture really is. More sympathetic readings consider it a benign caricature, promoting a simple but effective depiction of a complex, multifaceted enterprise. Crankier commentators dismiss it as an inadequate distortion that has overstayed its welcome. Still, few, if any, take it at face value, for good reason.

This being said, this superseded image is still very much alive in some circles. It is especially popular among the general public. Journalists, politicians, entrepreneurs, and many other non-specialists explicitly endorse and actively promote the vision of science as a gradual accumulation of facts and truths. While some tenants of the ivory tower of academia might take this as an opportunity to mock and chastise the incompetence of the masses, we should staunchly resist this temptation. First, note, we are talking about highly educated and influential portions of society. Lack of schooling can hardly be the main culprit. Second, and even more important, it is not hard to figure out where the source of the misconception might lie.

Textbooks, at all levels and intended for all audiences, present science as a bunch of laws, theories, facts, and notions to be internalized uncritically. This pernicious stereotype trickles down from schools and colleges to television shows, newspapers, and magazines, eventually reaching the general public. Now, surely, some promising young scholars will go on to attend graduate schools and pursue research-oriented careers: academic, clinical, governmental, and the like. They will soon learn, often the hard way, that actual scientific practice is very different—and way more interesting!—than what is crystallized in journals, books, and articles. Yet, this is a tiny fraction of the overall population. Most of us are only exposed to science in grade school or college, where the old view still dominates. By the time one leaves the classroom to walk into an office, the damage is typically done.

Where am I going with this? The bottom line is simple. The textbooks by which students learn science, which perpetuate the cumulative approach, are written by those same specialists who, in their research, eschew that brick-by-brick analogy. Why are experts mischaracterizing their own disciplines, promoting an old image that they themselves no longer endorse?

This book addresses this old question, which has troubled philosophers of science, at least since the debates of the 1970s in the wake of Kuhn's *Structure*. My answer can be broken down to a *pars destruens* and a *pars construens*. Beginning with the negative claim, textbooks do not adopt the cumulative model with the intention to deceive. The reason is that no viable alternative is available. The current scene is dominated by two competing models of science, neither of which supplants the "brick-by-brick" ideal. There must be more effective ways of popularizing research, exposing new generations to the fascinating world of science. The positive portion of my argument involves a constructive proposal. The next two sections present these two theses in embryonic form. Details will emerge in the ensuing chapters.

§1.2. A Theory of Everything?

The previous section concluded by posing a puzzle. Why do textbooks present science as an accumulation of truths and facts, given that their authors know perfectly well how inaccurate this is? The answer, I maintain, is that no better alternative is available. We currently have two competing models of science: reductionism and antireductionism. Neither provides an accurate depiction of the productive interaction between knowledge and ignorance, supplanting the old image of the wall. This section introduces the status quo, the two philosophical models presently dominating the scene.

Before getting started, two quick clarifications. First, these preliminary paragraphs merely set the stage for further discussion to be developed throughout the book. Skeptical readers unsatisfied with my cursory remarks are invited to hold on tight. Second, in claiming that "philosophical" models guide the presentation in textbooks and other popular venues, I am not suggesting that scientific pedagogy is directly responding to the work of philosophers—wishful thinking, one could say. Rather, my claim is that these models, developed systematically within the philosophy of science, reflect the implicit attitude underlying much empirical practice. With these qualifications in mind, we can finally get down to business.

Classic and contemporary philosophy of science has been molded by a
long-standing clash between two polarizing metaphysical stances. Allow me
to illustrate them by introducing a colorful image that will recur throughout
our discussion. Assume that the natural world can be partitioned into distinct, non-overlapping levels, arranged in terms of constitution, with smaller
constituents at the bottom and larger-scale entities arranged toward the top.
From this standpoint, these levels will correspond to traditional subdivisions
between fields or inquiries. Intuitively, physics, broadly construed, will be
at the very bottom. At a slightly coarser scale, we find entities described by
biology, followed by neuropsychology. Toward the top we will find the macroscopic structures postulated and studied by economics and other social sciences. It is customary to depict this layering of reality as a wedding cake. For
this reason, I shall refer to it as the "wedding cake model" (Figure 1.1). While
such partition will be further developed and refined, it is important to clarify
from the get-go that this representation is admittedly oversimplified and incomplete. It is incomplete because there are numerous fields and subfields of
science that are not included, such as sociology and anthropology. It is oversimplified because any student of science worth their salt is perfectly aware
that this partition does not capture complex, nuanced relations between disciplines. To wit, I am here clashing together psychology and neuroscience

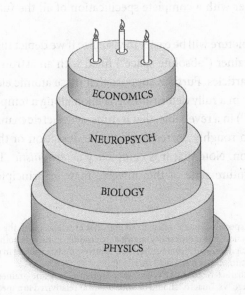

Figure 1.1. The "wedding cake" layering of reality.

under the moniker "neuropsychology," knowing perfectly well that their ontologies and methodologies diverge in important respects. Similarly, parts of biology overlap with psychology, neuroscience, and the social sciences, spawning hybrid, interdisciplinary approaches such as neuroethics, evolutionary psychology, and biolinguistics. Finally, one could contest my placement of physics at the very bottom, since a prominent part of this science, astrophysics, studies some of the largest systems in the universe.

With these limitations well in mind, I present this idealized model of reality to illustrate our two metaphysical stances. On the one hand, reductionism contends that all special sciences boil down to fundamental physics, in the sense that "higher levels" effectively reduce to the bottom level.[2] On the other hand, antireductionism maintains that "higher" layers are ontologically or epistemically autonomous from "lower" ones.[3] How should these claims be understood? What is the crux of the disagreement? A more precise articulation of the debate will be set aside until Chapter 2. For the time being, my aim is a general reconstruction of the main point of contention between reductionists and their foes.

Allow me to kick things off by introducing a simple thought experiment. Imagine that, at some time in the future, physics—or more accurately, the descendants of our current physical theories—will advance to the point of providing an exhaustive description of the bedrock structure of the entire cosmos, together with a complete specification of all the fundamental laws of nature.

At first, this picture will be easier to visualize if we depict the universe as an enormous container ("absolute space") filled with an astronomical number of indivisible particles. Further, suppose that these atomic elements interact with each other in a fully deterministic fashion along a temporal dimension ("absolute time") in a reversible, that is, time-symmetric manner.

This scenario roughly corresponds to the depiction of the cosmos provided by Newton. Note that it is completely *deterministic*. This means that any complete future state of the universe may, in principle, be predicted

---

[2] The expression "special sciences" refers to all branches of science, with the exception of fundamental (particle) physics. This moniker is somewhat misleading, as there is nothing inherently "special" about these disciplines, aside from having been relatively underexplored in philosophy. Yet, this label has become commonplace and I shall stick to it.

[3] To be clear, any characterization of levels as "coarse-grained" vs. "fine-grained," "higher-level" vs. "lower-level," or "micro" vs. "macro," should be understood as relativized to a specific choice of explanandum. From this standpoint, the same explanatory level L_n can be "micro" or "lower-level" relative to a coarser description L_{n+1}, and "macro" or "higher" relative to a finer-grained depiction L_{n-1}.

precisely from any comprehensive description of a past state, in conjunction with the relevant laws of nature. Analogously, past states can be retrodicted from future ones.

One final flight of fancy. Suppose that, down the line, researchers will also find a way to achieve the astounding computing power required for these derivations. What this means is that, given an exhaustive specification of the location and momentum of every bedrock particle in the universe, any past or future state can be determined with absolute certainty.

This thought experiment was popularized by the physicist and mathematician Pierre Simon de Laplace. For this reason, I shall follow an established tradition and refer to a creature with such stupendous predictive and retrodictive capacities as a "Laplacian Demon."[4]

As some readers will surely realize, making Laplace's insight cohere with contemporary physics requires some fancy footwork. First and foremost, we need a characterization of the universe that avoids any reference to the superseded concepts of absolute space and absolute time, and which reframes these notions in relativistic terms. Second, a Laplacian Demon worth their salt needs to be at least compatible with the fundamental laws of nature being indeterministic. This is a possibility left open by current quantum mechanics but, it should be noted, it is fundamentally at odds with one of Laplace's core tenets, namely, determinism. On this interpretation, the Demon will, at best, be able to assign to every past and future state a precise probability that falls short of unity, that is, certainty. Finally, a more realistic characterization of Laplace's thought experiment requires addressing thermodynamic and quantum irreversibility as well as in-principle unpredictable behaviors described by chaos theory.

Be that as it may, suppose that future physics follows this path, developing into a powerful framework capable of describing, down to its most basic constituents, every corner of the physical universe that exists, has existed, and will ever exist. These considerations raise a host of philosophical questions.[5] Under these hypothetical circumstances, what would happen to science as we know it today? Could physics explain every event? Or, more modestly, could it explain any event that currently falls under the domain

---

[4] For a discussion of the origins and foundations of Laplacian determinism, see van Strien (2014).

[5] "Philosophical" discussions of Laplacian Demons and the future of physics have been undertaken by prominent philosophers and scientists such as Dennett (1987); Putnam (1990); Weinberg (1992); Mayr (2004); Chalmers (2012); Nagel (2012); and Burge (2013).

of some branch of science? Would we still need the special sciences? Would physics replace them and truly become a scientific theory of everything?

There are two broad families of responses to these questions. The first, which may be dubbed *reductionism*, answers our quandaries in the positive. Reductionism comes in various degrees of strength. Any reductionist worth their salt is perfectly aware that current physics is still eons away from achieving the status of a bona fide accurate and exhaustive description of the universe. And even if our best available candidates for fundamental laws of nature happened to be exact, the computing power required to rephrase relatively simple higher-level events in physical terms remains out of reach, at least for now. This is to say that physics, as we presently have it, is not yet an overarching theory of everything. Still, the more radical reductionists claim, physics will eventually explain all events in the universe, and contemporary theories have already put us on the right track. More modest reductionists make further concessions. Perhaps physics will never actually develop to the point of becoming the utmost theory of reality. Even if it did, gaining the computing power to completely dispose of all non-physical explanations may remain a chimera. Hence, the special sciences will always be required in practice. Nevertheless, in principle, physics could do without them. In this sense, the special sciences are nothing but convenient, but potentially disposable, scaffoldings.

The second family of responses, *antireductionism*, provides diametrically opposite answers to our questions concerning the future of physics. Like its reductionist counterpart, antireductionism comes in degrees. More radical versions contend that, because of the fundamental disunity or heterogeneity of the universe, a physical theory of everything is a metaphysical impossibility. Even the most grandiose, all-encompassing, and successful physical theories could not explain all that goes on in the material universe because many—or, some would say, most—events covered by the special sciences fall outside of the domain of physics. Less uncompromising antireductionists may make some concessions toward modest forms of reductionism. Perhaps physics could, in principle, explain every scientific event. Still, the special sciences are nevertheless not disposable. This is because they provide objectively "better" explanations of higher-level happenings. In short, the antireductionist says, the success of physics is no threat for higher-level theories. Special sciences are not going anywhere, now or ever.

In sum, the debate between reductionism and antireductionism in general philosophy of science boils down to the prospects of developing a physical theory of everything, a grand characterization of the world that, in principle, could describe, predict, and explain any and every scientific event. These metaphysical considerations carry methodological implications. Are disciplines other than fundamental physics mere scaffoldings awaiting replacement? Or are special sciences a necessary, ineliminable component of any comprehensive description and explanation of the universe?

These thought-provoking questions have inspired much discussion. The divide between reductionism and antireductionism has dominated classic and contemporary philosophy of science. The widespread propensity to privilege fundamental physics over other sciences digs its roots deep into the history of *Naturwissenschaft*.[6] Logical positivism, the founding movement of the philosophy of science as it is currently understood, assumed that the physical and mathematical sciences set both the agenda and the tone of the conversation. This attitude that physics is the paradigm of true science and all other fields should sooner or later follow suit, still prominent in some circles, has attracted criticism among scientists, historians, and philosophers, spawning healthy debate. However, after decades of discussion, these positions have hardened, stifling productive conversation. The topic of reduction has attracted a disproportional amount of attention in philosophy, at the expense of other equally pressing and important issues. Let me motivate this claim, which lies at the core of my argument.[7]

First and foremost, debating whether all scientific inquiries can be addressed in terms of physics has a major drawback. It encourages laying all our eggs in one basket—the nature, limits, and boundaries of ideal, future science—setting aside questions pertaining to current inquiries. But how do we adjudicate, in the present, the prospects of research yet to come? In this respect, general philosophy of science has much to learn from contemporary philosophies of particular disciplines, which understood decades ago the importance of focusing on the here-and-now of empirical practice.

Second, and more relevant to our concerns, reductionism and antireductionism, as traditionally conceived, leave us wanting. One problem is that there is little consensus on how to characterize "reduction" and "autonomy," often leaving disputants talking past each other. In addition, neither stance is

---

[6] For an excellent discussion of the rise of scientific philosophy, see Friedman (2001).

[7] Similar conclusions have been reached, via a different route, by Wimsatt (2007), and developed by the "new wave of mechanistic philosophy," presented and examined in Chapter 7. Relatedly, Gillett (2016) has noted some discrepancy between the models of reduction and emergence developed in philosophy versus the ones adopted in scientific practice.

able to successfully capture the interplay between knowledge and ignorance in contemporary science. Reductionist metaphysics goes together well with the "brick-by-brick" epistemology, which views the advancement of science as breaking down complex, higher-level systems into smaller blocks. This leaves little room for productive ignorance, that is, the idea that ignorance itself may, at times, be preferable to the corresponding state of knowledge. Antireductionism, in contrast, legitimizes the role of ignorance by stressing the autonomy of the macrocosm. Still, it fails to convincingly explain why higher-level explanations can be objectively superior to lower-level ones. These two perspectives need to be combined, not pitched against each other.

Throughout the book, I shall argue that the reductionism vs. antireductionism opposition, as traditionally conceived, confronts us with a false dilemma. Paraphrasing Lakatos, despite its long-standing history and renewed popularity, this debate has now turned into a regressive research program. Both parties often talk past each other, rehashing old "this was never my point to begin with" claims. In the meantime, substantive historical, scientific, and philosophical questions accumulate, awaiting examination. After decades of discussion, it is finally time to move on. What we need is a framework that can bring together both aspects of science—autonomy and reduction—without pitching them against each other. What is needed, in short, is a different, alternative model of science. What could this be?

§1.3. Pandora's Gift to Science: The Black Box

Section 1.2 introduced my *pars destruens*: the debate between reductionism and antireductionism fails to provide an adequate image of the dynamic interplay between scientific knowledge and productive ignorance. If the current path is not the way to go, in what direction should we point the discussion?

Answering this question is the chief goal of the constructive portion of my argument. The chapters that follow sketch an account of scientific explanation that is neither reductionist nor antireductionist. Or, perhaps, it is both, in the sense that it combines core aspects from each perspective. The proposed shift brings attention to a prominent construct—the *black box*—which underlies a well-oiled technique for incorporating a productive role of ignorance and failure into the acquisition of empirical knowledge. While a comprehensive analysis will be undertaken throughout the monograph, let me briefly introduce you to this important albeit neglected concept.

Most readers will likely have some pre-theoretical, intuitive sense of what a black box is and may even have candidate examples in mind. If your acquaintance with electronics is anything like mine, your smartphone constitutes a perfectly good illustration, as it surely does for me. I am able to use my phone reasonably well. I can easily make calls, check my email, and look up driving directions. I am aware that, by opening a specific app, the screen will display my boarding pass, allowing me to board my flight. I obviously know that, if I do not charge the battery every other day, the darn thing will eventually run out of power. For all intents and purposes, I am a typical user. The point is that, like most customers, I have no clear conception of what happens inside the phone itself, as witnessed by the observation that, if something were to break or otherwise stop working, I would have to take it to a specialized store for repair. In brief, I have a fairly good grasp of the functional organization of my phone—systems of inputs and outputs that govern standard usage—and even a vague sense of the algorithms at play. But the inner workings that underlie this functionality and make it possible are a mystery to me. I am perfectly aware that they must be there. But, being no technician, I have no clue as to what exactly they are and how they work. In this sense, a smartphone is a black box to me, and I am confident that many others are in the same boat.

Do not let the mundane nature of the example deceive you. There is a long-standing historical tradition referring to "known unknowns," going back, at least, to the medieval debate on *insolubilia* and, more recently, Emil du Bois-Reymond's *ignoramus et ignorabimus*. In contemporary settings, references to black boxes can be found throughout the specialized literature across a variety of fields. In contemporary philosophy, the black box par excellence is the human mind itself. As Daniel Dennett (1991, p. 171) notes, in a passage of his book *Consciousness Explained* extracted from a section entitled "Inside the Black Box of Consciousness": "It is much easier to imagine the behavior (and behavioral implications) of a device you synthesize 'from the inside out' one might say, than to try to analyze the external behavior of a 'black box' and figure out what must be going on inside." The point is reinforced, from a slightly different perspective, in Pinker's *How the Mind Works* (1997, p. 4): "In a well-designed system, the components are black boxes that perform their function as if by magic. That is no less true of the mind. The faculty with which we ponder the world has no ability to peer inside itself or our other faculties to see what makes them tick." But consciousness is just the tip of the iceberg. References to black boxes can be found in the work of many prominent philosophers, such as Hanson (1963), Quine (1970), and Rorty (1979), just to pick a few notable examples.

And, of course, black boxes are hardly limited to the philosophy of mind, or even the field of philosophy *tout court*. As a contemporary biologist puts it, "the current state of scientific practice [. . .] more and more involves relying upon 'black box' methods in order to provide numerically based solutions to complex inference problems that cannot be solved analytically" (Orzack 2008, p. 102). And here is an evolutionary psychologist: "The optimality modeler's gambit is that evolved rules of thumb can mimic optimal behavior well enough not to disrupt the fit by much, so that they can be left as a black box" (Gigerenzer 2008, p. 55). These are just a few among many representative samples, which can be found across the board.

The use (and abuse) of black boxes is criticized as often as it is praised. Some neuroscientists scornfully dub the authors of functional models containing boxes, question marks, or other filler terms, "boxologists." In epidemiology— the branch of medicine dealing with the incidence, distribution, and possible control of diseases and other health factors—there is a recent effort to overcome the "black box methodology," that is, "the methodologic approach that ignores biology and thus treats all levels of the structure below that of the individual as one large opaque box not to be opened" (Weed 1998, p. 13). Many reductionists view black boxes as a necessary evil: something that does occur in science, but that is an embarrassment, not something to celebrate.

In short, without—yet—getting bogged down in details, references to black boxes, for better or for worse, are ubiquitous. Analogous remarks can be found across every field, from engineering to immunology, from neuroscience to machine learning, from analytic philosophy to ecology. What are we to make of these boxes that everyone seems to be talking about?

Familiar as it may ring, talk of boxes here is evidently a figure of speech. You may actually find a black box on an aircraft or a modern train. But you will not find any such thing in a philosophy professor's dusty office any more than you will find it in a library or research lab. What exactly is a black box? Simply put, it is a theoretical construct. It is a complex system whose structure is left mysterious to its users, or otherwise set aside. More precisely, the process of black-boxing a specific phenomenon involves isolating some of its core features, in such a way that they can be assumed without further micro-explanation or detailed description of its structure.

These issues will be developed throughout the monograph. Meanwhile, let me stress that not all black boxes are the same. Some work perfectly well. Others muddy the waters by hiding or obliterating crucial detail. Some boxes are opaque for everyone, as no one is currently able to open them. Others, like phones, depend on the subject in question. Some black boxes are constructed out of necessity: ignorance about the underlying mechanisms or computational intractability. Others are the product of error, laziness, or oversight. Yet again, some are constructed to unify fields. Still others draw disciplinary boundaries. In a nutshell, black-boxing is a more complex, nuanced, and powerful technique than is typically assumed.

Given these epistemic differences in aims and strategies, does it even make sense to talk about *the* practice of black-boxing, in the singular? This book is founded on the conviction that, yes, it does make sense. I shall argue that there is a methodological kernel that lies at the heart of all black boxes. This core phenomenon involves the identification of some entity, process, or mechanism that, for various reasons, can be idealized, abstracted away, and recast from a higher-level perspective without undermining its effectiveness, explanatory power, or autonomy. But how does this work? Answering these deceptively simple questions will require some effort.

In sum, here is our agenda. What is a black box? How does it work? How do we construct one? How do we determine what to include and what to leave out? What role do boxes play in contemporary scientific practice? If you have the patience to explore some fascinating episodes in the history of science, together we will address all of these issues. I shall argue that a detailed analysis of this widespread practice is descriptively accurate, in the sense that it captures some prominent albeit neglected aspects of scientific practice. It is also normatively adequate, in that it triggers a plausible analysis of the relation between explanations pitched in different fields, at different epistemic levels. Moreover, my proposal promises to bring together the insights on both reductionism and antireductionism, while avoiding the pitfalls of both extremes. Simply put, there are two aspects of ignorance in science. On the one hand, reductionism captures how ignorance points to what needs to be discovered. Antireductionism, on the other hand, reveals how science can proceed in the face of ignorance. Black boxes capture both dimensions, showing how autonomy and reduction are complementary, not incompatible, and offering a fresh perspective on stagnating debates. For all these reasons, black boxes are the perfect candidate for replacing stale models at the heart of philosophical discussions of scientific methodology.

This is our ambitious goal. Before getting down to business, we still have a couple of chores. First, I need to map the terrain ahead of us and clarify my perspective. This will be the task of section 1.4, which offers a synopsis of the chapters to come. Finally, section 1.5 concludes this preliminary overview with a couple of heads up and caveats about the aims and scope of the project.

§1.4. Structure and Synopsis

This book is divided into ten chapters, including the introduction you are currently reading. Chapter 2, "Between Scylla and Charybdis," provides an overview of the development of the reductionism vs. antireductionism debate, which has set the stage for philosophical analyses of science since the early decades of the twentieth century. Our point of departure is the rise and fall of the classical model of reduction, epitomized by the work of Ernest Nagel. Next is the subsequent forging of the "antireductionist consensus" and the "reductionist anti-consensus." Once the relevant background is set, the chapter concludes by arguing that modest reductionism and sophisticated antireductionism substantially overlap, making the dispute more terminological than is often appreciated. Even more problematically, friends and foes of reductionism tend to share an overly restrictive characterization of the interface between levels of explanation. Thus, it is time for philosophy to move away from these intertwining strands, which fail to capture the productive interplay between knowledge and ignorance in science, and to develop new categories for charting the nature and advancement of the scientific enterprise. Reductionism and antireductionism will return in the final chapter. Before doing so, we shall explore a new path by focusing on an explanatory strategy that, despite being well known and widely employed, currently lacks a systematic analysis. This strategy is black-boxing.

Chapter 3, "Lessons from the History of Science," starts cooking our alternative to "Scylla" and "Charybdis" by providing four historical illustrations of black boxes. The first two case studies originate from two intellectual giants in the field of biology. Darwin acknowledged the existence and significance of the mechanisms of inheritance. But he had no adequate proposal to offer. How could his explanations work so well, given that a crucial piece of the puzzle was missing? A similar shadow is cast on the work of Mendel and his early-twentieth-century followers, the so-called classical geneticists, who posited genes having little to no evidence of the nature, structure, or even the physical reality of these theoretical constructs. How can the thriving field of genetics be founded on such a fragile underpinning, a crackling layer of thin ice? The answer to both conundrums lies in the construction of black boxes, which effectively set to the side the unknown or mistaken details of these explanations without impacting their accuracy and robustness. Then came the Modern Synthesis, first, and the Developmental Synthesis later, which began to fill in the blanks, opening Darwin's and Mendel's black boxes, only to replace them with new black boxes. This phenomenon is by no means unique to biology. Another illustration is found in the elimination of mental states from the stimulus-response models advanced by psychological behaviorism. A final example comes from neoclassical economics, whose "as if" approach presupposes that the brain can be treated as a black box, essentially setting neuropsychological realism aside. The history of science, I shall argue, is essentially a history of black boxes.

In addition to illustrating the prominence of black boxes across the sciences, these episodes also show that, contrary to a common if tacit belief, black-boxing is hardly a monolithic, one-size-fits-all strategy that allows scientific research to proceed in the face of our ignorance. These theoretical constructs can play various subtly different roles. Yet, despite substantial methodological differences, there is a common thread. All four case histories point to the same core phenomenon: the identification of mechanisms that, for various reasons, are omitted from the relevant explanations. This is done via the construction of a black box. But what is a black box and how does black-boxing work? How are these entities constructed? What distinguishes a "good" box from a "bad" one? To answer these questions, we need to provide a more systematic analysis of this explanatory strategy.

Chapter 4, "Placeholders," poses the foundations of this project. It should be evident, even just from this cursory introduction, that black boxes function as placeholders. But what is a placeholder? What role does it play in science? I set out to answer these questions by introducing two widespread theses concerning the concept of biological fitness. First, fitness is commonly defined as a dispositional property. It is the propensity of an organism or trait to survive and reproduce in a particular environment. Second, since fitness supervenes—that is, depends, in a sense to be clarified—on its underlying physical properties, it is a placeholder for a deeper account that dispenses with the concept of fitness altogether. Plausible as they both are, these two theses are in tension. *Qua* placeholder, fitness is explanatory. *Qua* disposition, it explicates but cannot causally explain the associated behavior. In the second part of the chapter, I suggest a way out of this impasse. My solution, simply put, involves drawing a distinction between two kinds of placeholders. On the one hand, a placeholder may stand in for the range of events to be accounted for. In this case, the placeholder functions as a *frame*. It spells out an explanandum: a behavior, or range of behaviors, in need of explanation. On the other hand, a placeholder may stand in for the mechanisms, broadly construed, which bring about the patterns of behavior specified by the frame, regardless of how well their nature and structure are understood. When this occurs, the placeholder becomes an explanans and I refer to it as a *difference-maker*.

Both kinds of placeholders—frames and difference-makers—play a pivotal role in the construction of a black box. Chapter 5, "Black-Boxing 101," breaks down this process into three constitutive steps. First, in the *framing stage*, the explanandum is sharpened by placing the object of explanation in the appropriate context. This is typically accomplished by constructing a frame, a placeholder that stands in for patterns of behavior in need of explanation. Second, the *difference-making stage* provides a causal explanation of the framed explanandum. This involves identifying the relevant difference-makers, placeholders that stand in for the mechanisms producing these patterns. The final *representation stage* determines how these difference-makers should be portrayed, that is, which mechanistic components and activities should be explicitly represented, and which can be idealized or abstracted away. The outcome of this process is a model of the explanandum, a depiction of the relevant portion of the world. This analysis provides and justifies the general definition we were looking for. A black box is a placeholder—frame or difference-maker—in a causal explanation represented in a model. By now, we will be ready to put this to work.

Is this three-step recipe adequate and accurate? Does the proposed definition capture the essence of black-boxing? What are its advantages? What are its limitations? These questions are taken up in the following chapters. Chapter 6, "History of Science, Black-Boxing Style," revisits our case studies from the perspective of the present analysis of black boxes. By breaking down these episodes into our three main steps, we are able to see how it was possible for Darwin to provide a simple and elegant explanation of such a complex, overarching explanandum: distributions of organisms and traits across the globe. It also explains why Mendel is rightfully considered the founding father of genetics, despite having virtually no understanding of what genes are, how they work, and even if they existed from a physiological perspective.

Furthermore, if Darwin and Mendel are praised for skillfully setting the mechanisms of inheritance and variation aside and keeping them out of their explanations, why is Skinner criticized for providing essentially the same treatment of mental states? What distinguishes Darwin's and Mendel's pioneering insights from Skinner's influential, albeit outmoded, approach to psychology? Finally, our analysis of black boxes sheds light on the contemporary dispute over the goals and methodology of economics, dividing advocates of traditional neoclassical approaches from more or less revolutionary forms of contemporary "psycho-neural" economics.

After providing a systematic definition of black boxes and testing its adequacy against our case histories, it is time to address and enjoy the philosophical payoff of all this hard work. This begins in Chapter 7, "Diet Mechanistic Philosophy," which compares and contrasts the black-boxing approach with a movement that has gained much traction in the last couple of decades within the philosophy of science: the "new wave of mechanistic philosophy." The new mechanistic philosophy was also born as a reaction to the traditional reductionism vs. antireductionism divide. Unsurprisingly, it pioneers many of the theses discussed here. This raises a concern. Is my treatment of black boxes as novel and original as I claim? Or it is just a rehashing of ideas that have been on the table since the turn of the new millennium? As we shall see, the black-boxing recipe fits in quite well with the depiction of science being in the business of discovering and modeling mechanisms. All three steps underlying the construction of a black box have been stressed, in some form or degree, in the extant literature. Nevertheless, the construction of black boxes, as I present it here, dampens many of the ontological implications that characterize the contemporary landscape. This allows us to respond to some objections raised against traditional mechanism. For this reason, I provocatively refer to black-boxing as a "diet" mechanistic philosophy, with all the epistemic flavor of your ole fashioned mechanistic view, but hardly any metaphysical calories. Now, we can really begin to explore philosophy of science, black-boxing style.

Reductionism contends that science invariably advances by descending to lower levels. Antireductionism flatly rejects this tenet. Some explanations, it claims, cannot be enhanced by breaking them down further. But why should this be so? What makes explanations "autonomous"? A popular way of cashing out the antireductionist thesis involves the concept of *emergence*. The core intuition underlying emergence is simple. As systems become increasingly complex, they begin to display properties which, in some sense, transcend the properties of their parts. The main task of a philosophical analysis of emergence is to cash out this "in some sense" qualifier. In what ways, if any, do emergents transcend aggregative properties of their constituents? How should one understand the alleged unpredictability, non-explainability, or irreducibility of the resulting behavior? Answering these questions might seem simple at first glance. But it has challenged scientists and philosophers alike for a hot minute. Chapter 8, "Emergence Reframed," presents, motivates, and defends a strategy for characterizing emergence and its role in scientific research, grounded in our analysis of black boxes. Emergents, I maintain, can be characterized as black boxes: placeholders in causal explanations represented in models. My proposal has the welcome implications of bringing together various usages of emergence across domains and to reconcile emergence with reduction. Yet, this does come at a cost. It requires abandoning a rigid perspective according to which emergence is an intrinsic or absolute feature of systems, in favor or a more contextual approach that relativizes the emergent status of a property or behavior to a specific explanatory frame of reference.

Chapter 9, "The Fuel of Scientific Progress," addresses a classic topic that, over the last couple of decades, has been unduly neglected: the question of the advancement of science. Setting up the discussion will require us to retrace our steps back to the roots of modern philosophy of science. Logical positivism provided an intuitive and prima facie compelling account of scientific knowledge. Science advances through a slow, constant, painstaking accumulation of facts or, more modestly, increasingly precise approximations thereof, in a "brick-by-brick" fashion (§1.1). These good old days are gone. In the wake of Kuhn's groundbreaking work, positivist philosophy of science was replaced by a more realistic and historically informed depiction of scientific theory and practice. However, over half a century has now passed since the publication of *Structure*. Despite valiant attempts, we still lack a fully developed, viable replacement for the cumulative model presupposed by positivism. At the dawn of the new millennium, mainstream philosophy eventually abandoned the project of developing a grand, overarching account of science. The quest for generality was traded in for a more detailed analysis of particular disciplines and practices. I shall not attempt here a systematic development of a post-Kuhnian alternative to logical positivism. More modestly, my goal is to show how the black-boxing strategy can offer a revamped formulation of scientific progress, an important topic that lies at the core of any general characterization of science, and to bring it back on the philosophical main stage, where it legitimately belongs.

Chapter 10, "Sailing through the Strait," takes us right back to where we started. Chapter 2 characterizes contemporary philosophy of science as metaphorically navigating between Scylla and Charybdis, that is, between reductionism and antireductionism. There, I ask two related families of questions. First, is it possible to steer clear of both hazards? Is there an alternative model of the nature and advancement of science that avoids the pitfalls of both stances and, in doing so, provides a fresh way of presenting science to an educated readership in a more realistic fashion? Second, how does science bring together the productive role of ignorance and the progressive growth of knowledge? The final chapter cashes out these two promissory notes. These two sets of problems have a common answer: black boxes. Specifically, the first four sections argue that the black-boxing strategy outlined throughout the book captures the advantages of both reductionism and antireductionism, while eschewing more troublesome implications. The final section addresses the interplay of ignorance and knowledge.

§1.5. Aim and Scope

I conclude this introductory overview by clarifying the aim and scope of this work. At the most general level, I have three main targets in mind.

My first goal is a philosophical analysis of an important theoretical construct and how it affects scientific practice. Talk about black boxes is ubiquitous. This metaphor is widely employed by scientists, philosophers, historians, sociologists, politicians, and many others. Yet, no one ever tells us exactly what to make of this figure of speech. I offer to pick up this tab. I should make it clear that my aim is hardly to dismiss and replace all the excellent work in general history and philosophy of science that has been developed over the last few decades. The collective goal of the ensuing chapters is twofold. On the one hand, they develop and refine the process of black-boxing by appealing to some traditional debates in general philosophy of science. On the other hand, I also want to suggest that black boxes provide a clear and precise framework to systematize an array of traditional concepts, whose nature has proven to be notoriously elusive, especially after the unifying force of logical positivism has waned. More generally, my objective is to advocate and justify a shift in perspective. If we move away from the

traditional reductionism vs. antireductionism divide and recast these important questions in terms of black boxes, we shed new light on these projects.
Admittedly, this is an ambitious endeavor. Thus, I must convince you that
this bold move is worth the cost, by showing the potential payoff in the form
of many new exciting research projects.

My second goal is, broadly speaking, historical. I maintain that black
boxes play a prominent role in the development and advancement of science.
For my arguments to be persuasive, the case studies need to be both accurate
and illuminating. Admittedly, this book is not, primarily, a scholarly contribution to the history of science—an endeavor that lies beyond my professional competence. My project falls more within the confines of a rational
reconstruction, as traditionally conceived in the philosophy of science.[8] At
the same time, this work is also intended to have some historical significance,
as black boxes are important to understand not merely what these authors
should have done, but also what they actually said and did.[9]

My third goal pertains more directly to the philosophy of science. As we
shall see in the ensuing chapters, black-boxing provides the intellectual tools
to understand and advance various foundational concepts. My objective is
to begin revisiting questions about causation, explanation, emergence, progress, and the old reductionism vs. antireductionism debate in a new guise.
This requires a blend of conceptual rigor and historical accuracy.

Finally, a word about my intended readership. This book should be of interest to both philosophers and theoretically inclined scientists, as well as
students of each discipline. Beginning and advanced scholars should have
no problem navigating the arguments and concepts presented here. The exposition is entirely self-contained, with the exception of some basic formal
logic. My hope, in keeping the discussion widely accessible, is to encourage
researchers from various disciplines to venture into the depths of this exciting topic.

---

[8] The notion of rational reconstruction is rooted in the groundbreaking work of Carnap (1956b);
Kuhn (1962); and, later, Lakatos (1976).

[9] I try to adhere to two adequacy conditions borrowed from Kitcher (1993, pp. 12–13). First, if
something is attributed to a figure, that attribution is correct. Second, nothing is omitted which, if
introduced into the account, would undermine the point made.

2

Between Scylla and Charybdis

Come fa l'onda là sovra Cariddi,
che si frange con quella in cui s'intoppa,
così convien che qui la gente riddi.

—Dante, Inferno, *VII*, 16–18[*]

§2.1. Introduction

Our discussion began, in Chapter 1, with a well-known image of science. From this perspective, the scientific enterprise is properly conceived as a progressive "brick-by-brick" accumulation of knowledge and facts. Despite its hallowed history and familiar ring, this view is now rejected by most specialists. Yet it remains popular among the educated general public, who take it straight out of our textbooks. These considerations raise a question. Why are we actively promoting such a distorted characterization of science?

This book constitutes a long answer to this query, together with a sketch of an alternative solution. The constructive proposal will be spelled out in the ensuing chapters. Before doing so, we need to address the status quo. My *pars destruens* can be presented as follows. We currently have two general models that implicitly guide much scientific practice, and which have been developed, more explicitly, within theoretical and philosophical discussions of scientific methodology. Neither stance is fully satisfactory. Overcoming the old cumulative image while, at the same time, explaining the productive role of ignorance requires us to move past extant conceptualizations. Motivating this claim is our present task. To get the proverbial ball rolling, I introduce a metaphor that will prove useful along the way.

---

[*] "Even as waves that break above Charybdis, // each shattering the other when they meet, // so must the spirits here dance the round dance." Translation by A. Mandelbaum.

Black Boxes. Marco J. Nathan, Oxford University Press. © Oxford University Press 2021.
DOI: 10.1093/oso/9780190095482.003.0002

§2.2. The Rock Shoal and the Whirlpool

In Greek mythology, sailors who dared to navigate across the Strait of Messina were confronted by two deadly perils, positioned so close to each other to pose an inescapable trap. The first threat was Scylla, a rock shoal located along the Italian shore, colorfully portrayed by Homer as a six-headed monster. The second was Charybdis, a huge whirlpool off the island of Sicily. While the beautiful Mediterranean coastline has long lost this poetic aura of mystery, the image has stuck. Being "between Scylla and Charybdis" figuratively refers to a forced choice between one of two evils.

This metaphor portrays the condition of much contemporary philosophy of science. Philosophers are currently "sailing" through two conceptual hazards: the *Scylla of reductionism* and the *Charybdis of antireductionism*. Let me begin by presenting my thesis in embryonic form. These claims are unraveled, in greater depth, in this chapter and throughout the book.

During the heyday of logical positivism, the received view was that all of science would eventually boil down to fundamental physics.[1] To be sure, logical empiricists were perfectly aware of how far science was from accomplishing even a distal approximation of such a complete derivation. Furthermore, positivists did not overlook how translating all scientific inquiries into questions for physics would pose practical hurdles for cognitively limited creatures like us. Predicting fluctuations in the stock market is challenging enough when financial transactions are understood in economic terms. Approaching them as interactions between subatomic particles would not make it any easier. Nevertheless, the in-principle feasibility of these reductions, together with their unificatory and guiding roles, were central tenets of positivist philosophy of science. This regulative ideal was conceived along the lines of what came to be known as the "classical model," which views reduction as a logical relation between theories.

Influential as it once was, the classical model has fallen on hard times. As we shall see, the positivist ideal was gradually eroded by powerful objections, fueled by the multiple-realizability of higher-level kinds, the lack of appropriate law-like generalizations, and the shortcomings of "syntactic" conceptions of theories. Consequently, throughout the 1970s and the 1980s,

---

[1] Some authors draw a fine-grained distinction between "logical positivism" and "logical empiricism," separating the early stages of the movement, revolving around the Vienna Circle, from later phases, following the post–World War II diaspora. Unless specified otherwise, for the sake of simplicity, I shall use the two expressions interchangeably.

an "antireductionist consensus" was forged. On this view, progressive reduction is doomed to fail, both normatively and descriptively, as an account of science.

Unfortunately, it turns out that antireductionism itself is not devoid of controversy. When advanced as a negative thesis—the in-principle impossibility of reducing all of science to physics—it leaves us without a viable methodological alternative. When cashed out as a constructive proposal, most antireductionist projects are confronted with an array of nagging concerns. Is antireductionism consistent with a modest materialist ontology, which accepts the supervenience of all scientific entities on physical properties? Can antireductionism, which purports to vindicate the "autonomy" of higher sciences, explain the thriving of interdisciplinary projects, which successfully borrow concepts and merge theories from various disciplines?

Because of these and related worries, the last few decades have witnessed the rise of a "reductionist anti-consensus" across the philosophy of science. Recent approaches steer clear of the logical derivations which characterized earlier forms of reductionism, replacing them with revamped epistemic frameworks. The general contention fueling the new wave of reductionism is that scientific explanations are always empowered by shifting down to lower levels, that is, by providing descriptions of the explanandum at smaller scales. While contemporary reductionist models avoid many of the problems that plagued their positivist predecessors, they have attracted a brand new host of objections. Antireductionists have new targets in sight.

Over half a century after the demise of logical positivism, the battle over the nature and status of reductionism rages on. While no winners or losers are in sight, both armies look fairly worn out. Each side strenuously continues its fight, constantly refining its arguments and offering sharper rebuttals of opposing frameworks. But is this the best strategy for heading forward? Might it not be time to call for a truce and reshuffle the deck?

This chapter argues that the reductionism vs. antireductionism rhetoric poses a false dilemma. After surveying, in greater detail, the unfolding of the debate, I suggest that sophisticated versions of both stances overlap, making the opposition more verbal and less substantive than many disputants like to admit. The root of the problem is that reductionists and antireductionists alike share a restrictive interface between levels of explanations, viewing reduction and autonomy as antithetical, as opposed to conceiving them as two sides of the same coin. Thus, both stances provide an impoverished conception of the nature and advancement of science.

If this old-fashioned debate is not a promising way forward, in what directions should we point our research? The goal of this book is to trace a different route, thereby suggesting a recharacterization of the current philosophy of science agenda. My proposal focuses on the role of black boxes in scientific explanation. These epistemic constructs identify entities, processes, and mechanisms that, for a variety of reasons, may be idealized or abstracted away from our models without any significant loss of explanatory power. These black boxes shed light on the productive side of ignorance.

The following chapters provide a systematic analysis of black-boxing and illustrate the general theme by presenting some prominent episodes in the history of science. We shall return to the present topic—the standoff between reductionism and antireductionism—in Chapter 10. Specifically, I shall then make the case that black-boxing captures insights common to both modest reductionism and sophisticated antireductionism, while eschewing problems and accommodating ignorance. Yet, before embarking on these ambitious quests, allow me first to convince you that it is time to sail away from both Scylla and Charybdis, in search of calmer waters.

Before moving on, a fair warning is due. The present chapter may feel a little technical, especially for readers unacquainted with the relevant literature. However, the details are crucial to appreciate the character and limitations of the current debate. Scholars who are already familiar with the issues under discussion should consider moving directly to section 2.6. Empirically oriented readers with little patience for philosophical analysis are advised to skip this chapter entirely and begin with Chapter 3.

§2.3. The Rise and Fall of Classical Reductionism

Our excursus into the modern history of reductionism sets out by going back to the dawn of the contemporary debate. The story starts with the classical model of derivational reduction, a legacy of logical positivism.[2]

In what came to be a locus classicus, Ernest Nagel (1961) characterized reduction as the logical deduction of the laws of a reduced theory S from the laws of a reducing theory P, a condition known as "derivability."[3] For such

[2] The exposition in this section draws heavily from Nathan and Del Pinal (2016).
[3] "S" and "P" stand in, respectively, for one of the special sciences and fundamental physics. Yet, the same schema can be applied, *mutatis mutandis*, to any pair of sciences hierarchically arranged: neuroscience and psychology, biology and chemistry, economics and sociology, etc. For the sake of simplicity, let us assume that the language of these theories does not substantially overlap, that is, most

derivation to get off the ground, a second condition of "connectability" must also be met. It has to be possible for the predicates figuring in the laws of S to be expressed in terms of predicates belonging to the language of P. To wit, suppose that among the laws of S there is the generalization L_S: $S_1x \rightarrow S_2x$, which connects property S_1x with property S_2x in a nomological fashion. To reduce this law to theory P, on Nagel's model, one needs to show that L_S can be derived from a law L_P: $P_1x \rightarrow P_2x$, expressed entirely in the language of theory P. What is needed is a series of *bridge laws*, principles that govern the systematic translation of the relevant S-predicates into P-predicates. Such principles will have the following form:

$$(R_1) \ S_1x \leftrightarrow P_1x$$
$$(R_2) \ S_2x \leftrightarrow P_2x$$

How should the biconditional be interpreted, for bridge principles like R_1 and R_2 to fulfill their role in the context of Nagel-style reduction?

First, "$\leftrightarrow$" must be transitive. This ensures that, if S_1 is reduced to Q_1, and Q_1 is reduced to P_1, then S_1 is thereby reduced to P_1. This proviso allows the reduction of any branch of science to physics to occur in a stepwise fashion. Thus, one need not derive, say, economics from biology in a single swoop. Such operation may be enacted gradually by, first, reducing economics to psychology and, subsequently, reducing psychology to biology.

Second, and more important, one should not read "$S \leftrightarrow P$" as "S causes P." The reason is that causal relations are typically asymmetric. Causes bring about their effects, but effects usually do not bring about their causes. Bridge principles, in contrast, must be symmetric. If an S_1-event is a P_1- event, then a P_1-event is also an S_1-event. Given these two requirements, bridge laws are most naturally interpreted as expressing *contingent event identities*. Thus understood, R_1 states that S_1 is type-identical to P_1.[4] To illustrate, a lightning-event is the same type of event as a discharge-of-electricity-in-the-atmosphere

predicates of S are not predicates of P and, vice versa, the majority of P-predicates do not belong to the vocabulary of S.

---

[4] As Fodor (1974) notes, the bridge laws required by Nagelian reduction express a stronger position than *token physicalism*, the plausible view that all events which fall under the laws of a special science supervene on physical events. Statements like R_1 and R_2 presuppose *type physicalism*, a more controversial tenet according to which kinds figuring in the laws of special sciences must be type-identical to more fundamental kinds.

event. Along the same lines, tasting sea salt and tasting sodium chloride (*NaCl*) are identical types of events.

Intuitive and influential as it once was, Nagel's framework has now fallen on hard times, undermined by powerful methodological arguments. Before presenting some objections, it is instructive to focus on a couple of desiderata that the classical model accomplishes and does so quite well.

First, reductionism, in general, offers a clear and precise concept of the unity of science. Two theories are said to be unified when one is reduced to the other or both are subsumed under a broader, more general theory. To be sure, many contemporary scholars reject the identification of unity with derivational reduction. Some even question whether science should be viewed as unified at all.[5] Yet, this is typically a side effect of the shortcomings of reductionism, specifically, of Nagel's model. To the extent that reduction offers a viable general model of science, it also captures its unity. Thus, unsurprisingly, during the heyday of positivism, the unity of science was treated as a matter of logical relations between the terms and laws of various fields, to be achieved through a series of inter-theoretic reductions.[6]

A second accomplishment of classical reductionism is the provision of a clear-cut account of how lower-level discoveries can, in principle, inform higher-level theories. Let me illustrate the point with simple examples. Suppose that we are testing a psychological hypothesis L_{Psy}: $Psy_1 x \rightarrow Psy_2 x$, which posits a law-like connection between two psychological predicates: Psy_1 and Psy_2. If we had a pair of reductive bridge laws that map Psy_1 and Psy_2 onto neural kinds Neu_1 and Neu_2, then we could confirm and explain the nomological status of L_{Psy} directly by uncovering the neural-level connection L_{Neu}: $Neu_1 x \rightarrow Neu_2 x$. This is because, as noted, the bridge laws presupposed by derivational reduction express contingent type-identities. If Psy_1 and Psy_2 are type-identical to Neu_1 and Neu_2, and there is a law-like connection between Psy_1 and Psy_2, there will also be a corresponding nomological connection between Neu_1 and Neu_2.

Readers unfamiliar with these debates might find the following analogy more intuitive. Assume that *water* and *sea salt* are type-identical to H_2O and *NaCl*, respectively. If one provided a successful explanation of why *NaCl* dissolves in H_2O, under specific circumstances, then one has thereby explained why sea salt dissolves in water, under those same conditions.

---

[5] Prominent examples are Dupré (1993) and Cartwright (1999).
[6] This tenet is made explicit in Carnap (1938) and Oppenheim and Putnam (1958).

In short, the reductive approach suggests a general model for how micro-theories can be used to advance their macro-counterparts. The goal is to look for lower-level implementations of higher-level processes, which—on the presupposition of reductionism—can then be used directly to test and explain macroscopic laws and generalizations. Reduction is unification, and unification, the story goes, is the goal and benchmark of scientific progress.

The good news is over. Now, on to some objections. The best-known problem with derivational reduction stems from the observation that natural kinds seldom correspond neatly across levels in the way presupposed and required by reductive bridge laws. One could arguably find a handful of successful Nagelian reductions in the history of science. For instance, one could make a strong case that sea salt and $NaCl$ are type-identical, that the action-potential of neurons is derivable from electric impulses, or that heat has been effectively reduced to the mean kinetic energy of constituent molecules. Assume, for the sake of the argument, that these reductions do, in fact, fit in well with the classical model presented earlier.[7] Still, contingent event identities are way too scarce to make classical reductionism a plausible, accurate, and general inter-theoretic model of scientific practice. In most cases, there are no physical, chemical, or macromolecular kinds that correspond—in the sense of being type-identical—to biological, psychological, or economic kinds, in the manner required by Nagel's framework. This, in a nutshell, is the multiple-realizability argument, first spelled out by Putnam and Fodor. The basic idea is that most higher-level kinds are multiply-realizable and functionally describable. Consequently, we rarely have bridge principles like R_1 and R_2 ("$S_1 x \leftrightarrow P_1 x$"; "$S_2 x \leftrightarrow P_2 x$"), which posit one-to-one mappings of kinds across levels. Rather, what we typically find in science are linking laws such as R_3, which capture how higher-level kinds can be potentially realized by a variety of lower-level states:

$$(R_3)\; S_1 x \leftrightarrow P_1 x \vee \ldots \vee P_n x$$

For most higher-level predicates S, the number of corresponding lower-level predicates P is likely to be quite large. To make things worse, all the P_i will typically have little to nothing in common, from a lower-level perspective.

---

[7] This is hardly uncontroversial. For some difficulties and qualifications affecting the "classical" reduction of thermodynamics to statistical mechanics, see Sklar (1993).

A second difficulty with classical reductionism concerns the underlying conception of theories. Recall that, on Nagel's model, the relata in a reductive relation are theories. On the standard post-positivist, "syntactic" interpretation, theories are conceived as collections of statements, including both empirical laws and testable conclusions derivable from them. Note how well this approach, adopted by many introductory scientific textbooks, fits in with the "brick-by-brick" conception presented in Chapter 1. However, as we shall see in section 2.4, the complex structure of many areas of the special sciences cannot be captured in terms of laws and law-like generalizations, on pain of generating a dramatically impoverished reconstruction.

There are two standard responses to this worry. One is to revise the definition of "theory." For example, one could replace the syntactic conception with a *semantic* approach that identifies scientific theories with collections of models.[8] Alternatively, one can deny that genetics, biochemistry, and evolution count as theories at all. "Theory" can then be substituted with an altogether different term, such as Kuhn's "paradigms," Toulmin's "disciplines," Laudan's "research programs," or Shapere's "fields." Whether scientific subfields are better defined as "theories" or as something else is a question that transcends the scope of our discussion. The relevant point is that neither strategy will salvage classical reduction. No matter what terms we use to define them, the special sciences cannot be adequately captured and described as the interpreted axiomatic systems presupposed by Nagel's model. This hammers another nail in the coffin of derivational reduction.

Trouble is not quite over. Setting aside the problem of multiple realizability and the limitations of interpreted axiomatic systems, derivational reduction has also been dubbed inadequate as a characterization of current trends in scientific practice. As noted, classical reductionism identifies scientific progress with successful reduction. Hence, it carries the presupposition that, with the advancement of the scientific enterprise, we should witness a recurring reduction of higher-level fields to lower-level ones. Nevertheless, this prediction has not been borne out. As stressed by many, the proliferation of new fields is as frequent, or even more common, than their merger and reduction. Moreover, with the possible exception of a few hackneyed and controversial examples, clear-cut instances of reduction are hard to come by. The situation is especially troublesome in the context of biology, where even the best candidates for Nagelian derivation, namely, Mendelian and molecular

---

[8] This view is spelled out in Suppes (1960); van Fraassen (1980); and Lloyd (1988).

genetics, notoriously fail (§2.4). Finally, classical reductionism is not only seemingly descriptively inaccurate. Even its normative force as a regulative ideal, as a goal for science, has been questioned. Derivational reduction, it is claimed, leaves unanswered—indeed, unasked—important questions about interactions between branches of science, intermediate or linking theories, and the way the unity of science actually comes about.[9]

In sum, classical theoretical reductionism has fallen on all fronts. The multiple realizability of kinds poses an apparently insurmountable challenge. The traditional syntactic conception of theories is drastically limited in scope. And the framework lacks both normative and descriptive force.[10] To be sure, as we shall see in section 2.5, reductionism remains a popular stance, with feisty defenders across the philosophy of science. Yet, contemporary reductionism has morphed. Current debates focus on epistemic questions concerning whether higher-level events are always best explained at more fundamental levels, not on the logical derivation of laws across science.[11] Theory reduction, in its traditional guise, has been beaten to death.

§2.4. Antireductionism Strikes Back

If antireductionism were tantamount to the rejection of Nagel's model of derivational reduction, then it would be virtually uncontroversial. However, antireductionists do not merely advance a negative tenet. Their *pars destruens* is supplemented with a *pars construens*. The general positive thesis can be characterized as the explanatory autonomy of the higher levels. This section surveys some influential arguments which attempt to

---

[9] For an excellent discussion of these points, see Maull (1977).

[10] This list of objections is by no means exhaustive. Other influential difficulties with Nagel's original model include an incompatibility with various forms of meaning holism and the observation that, when a theory succeeds another in an area of science, the theories typically make contradictory claims (Hütteman and Love 2016).

[11] There have been notable attempts to amend Nagel's framework. Schaffner (1967, 1993) argues that the target of the deduction from the lower-level theory is not the original higher-level theory, but a suitably corrected version. More recently, Dizadji-Bahmani et al. (2010) have generalized Schaffner's approach by further restricting the lower-level theory with boundary conditions and auxiliary assumptions. While these refinements constitute significant improvements, they admittedly fail to provide a general model of reduction that holds across the sciences. Following a different strategy, other authors have attempted to rehabilitate Nagel by arguing that traditional efforts to apply derivational reduction misinterpret his "condition of connectability" (Fazekas 2009; Klein 2009). The prospects of salvaging Nagel's model by reinterpreting it constitute an interesting project, albeit one that I shall not address here. My point is simply that classical derivational reduction, in its original, most popular, and overarching form, is doomed.

establish this epistemic claim.[12] Next, section 2.5 will consider some reductionist comebacks.

One of the most influential arguments in support of explanatory autonomy was originally developed by Putnam in his article "Philosophy and Our Mental Life." Putnam's intended goal was to show that traditional philosophical discussions of the mind-body problem rest on a misleading assumption. He begins by noting how all parties involved presuppose, more or less implicitly, the following premise. If human beings are purely material entities, then there must be a physical explanation of our behavior. Materialists employ this conditional claim in a modus ponens inference. Together with the additional premise that we are material entities, it vindicates the in-principle possibility of physically explaining our behavior:

(a) If humans are purely material entities, then there is a physical explanation of human behavior.
(b) Humans are purely material entities.
(c) ∴ There is a physical explanation of human behavior.

Dualists accept premise (a) but turn the modus ponens into a modus tollens by revising (b) and (c). Given that there can be no physical explanation of human behavior, they argue, we cannot be purely material beings:

(a) If humans are purely material entities, then there is a physical explanation of human behavior.
(c*) There is no physical explanation of human behavior.
(b*) ∴ Humans are not purely material entities.

[12] It is important to distinguish the *epistemic* form of antireductionism addressed here from various *metaphysical* variants. For instance, Galison, Hacking, Cartwright, Dupré, and other members of the "Stanford School" have defended an ontological thesis, the heterogeneity of the natural world, from which follows a methodological tenet fundamentally at odds with the positivist outlook: the disunity of science. Reductionism, which Dupré (1993, p. 88) defines as "the view that the ultimate scientific understanding of phenomena is to be gained exclusively from looking at the constituents of those phenomena and their properties" becomes a derivative target, in virtue of its connection with the unity of science. A different form of metaphysical antireductionism, associated with the possibility of downward causation and strong emergence, has become popular in science in the context of systems biology. We shall discuss some of these metaphysical stances—in particular, emergence—in Chapter 8. For the moment, I shall only be concerned with antireductionism understood primarily as an epistemic tenet of scientific explanation.

Both arguments, Putnam contends, miss the mark. Physicalists and dualists make the same mistake by accepting the conditional premise (a). To establish his point, he introduces an example that has since become famous.

Consider a simple physical system constituted by a rigid board with two holes—a circle one inch in diameter and a square one inch high—and a rigid cubical peg just under one inch high (Figure 2.1). Our task is to explain the intuitive observation that the peg can pass though the square hole, but it will not go through the round hole. Why is this the case?

Putnam sketches two kinds of explanations. The first begins by noting that both the board and the peg are rigid lattices of atoms. Now, if we compute the astronomical number of all possible trajectories of the peg, we will discover that no trajectory passes through the round hole. There is, however, at least one (and likely quite a few) trajectories that will pass through the square hole. A second kind of explanation begins in exactly the same way, by noting that the board and the peg are both rigid systems. But, rather than computing trajectories, it points out that the square hole is slightly bigger than the cross-section of the peg, whereas the round hole is smaller. Let us call the first kind of explanation "lower-level," "micro," or "physical" and, correspondingly, label the second one "higher-level," "macro," or "geometrical." The predictable follow-up question is: are both explanations adequate? If not, why not? And, if so, which one is better?

Putnam maintains that the geometrical explanation is objectively superior, from a methodological perspective, to its physical counterpart. (Actually, he goes as far as saying that the physical description is not an explanation at all.

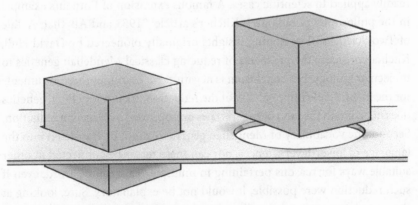

Figure 2.1. Putnam's square-peg-round-hole scenario.

Yet, we can set this more controversial thesis to the side.) The reason is that the physical description evidently applies only to the specific case at hand, since no other pegs and boards will have exactly the same atomic structure. In contrast, the geometrical story can be straightforwardly generalized to similar systems. Whereas the macro-explanation brings out the relevant geometrical relations, the micro-explanation conceals these laws. As Putnam puts it: "in terms of real life disciplines, real life ways of slicing up scientific problems, the higher-level explanation is far more general, which is why it is *explanatory*" (1975a, p. 297).

The moral drawn by Putnam from his example is the explanatory autonomy of the mental from the physical. Higher-level explanations, regardless of whether they involve pegs or mental states, should not be explained at lower levels, in terms of biochemical or physical properties. Doing so does not produce a better explanation of the explanandum under scrutiny.

Putnam's conclusion has fired up a long-standing debate that rages on to the present day. We can all agree that, for creatures like us, the geometrical explanation is simple and good enough for most intents and purposes. Furthermore, the physical explanation is likely intractable and overkill. That much virtually no one disputes. But is it really true that macro-explanations are objectively superior to their lower-level counterparts, in the sense that, if we were to advance them, they would explain more and better? As we shall see in section 2.5, reductionists beg to disagree. Before getting there, in the remainder of this section, let us focus on how Putnam's insight has contributed to forging an antireductionist consensus.

The square-peg–round-hole analogy involves a simple toy system. Yet, its general conclusion, the explanatory autonomy of the macro-level, can be readily applied to scientific cases. A famous extension of Putnam's example in the philosophy of biology is Kitcher's article: "1953 and All That: A Tale of Two Sciences." Developing insights originally pioneered by David Hull, Kitcher considers the prospects of reducing classical Mendelian genetics to molecular biology. His conclusion is negative. He examines three arguments for the in-principle impossibility of the reduction in question. First, genetics does not contain the kind of general laws presupposed by Nagelian reduction. Second, the vocabulary of Mendelian genetics cannot be translated into the language of lower-level sciences, nor can these terms be connected in other suitable ways for reasons pertaining to multiple-realizability. Third, even if such reduction were possible, it would not be explanatory. Sure, looking at the underlying mechanisms allows one to extend higher-level depictions in various ways. Nevertheless, appeals to molecular biology do not deepen cytological explanations of genetic phenomena.

While all three points are worthy of discussion, I focus mainly on the third one, for two basic reasons. First, Kitcher's final objection is directly connected to Putnam's square-peg–round-hole example. Second, and more important, the first two claims are, nowadays, seldom contested. Even the staunchest contemporary reductionist would concede the hopelessness of reducing Mendelian to molecular genetics along the lines of Nagel's model. This is because, as Kitcher aptly notes, we lack the laws and bridge principles required for applying the framework in question. In contrast, as we shall see in the following, current reductionism takes issue with the third argument.

To establish his conclusion, Kitcher considers a modern paraphrase of Mendel's second law, the "law of independent assortment," stating that genes on non-homologous chromosomes assort independently. How does one explain this generalization? Biology textbooks typically do so by providing a description of the cytological process of *meiosis* at the level of cell bodies and their interactions. The standard story goes something like this. Chromosomes line up with their homologues, making it possible for them to exchange genetic material and produce pairs of recombinant chromosomes. In the course of this process, one member of each recombinant pair goes to each gamete in a way that makes the assignment of a member of another pair to that same gamete statistically independent. It follows that genes on homologous chromosomes will likely be transmitted together, whereas genes on non-homologous chromosomes tend to assort independently.

Just like Putnam, Kitcher considers the prospects of explaining Mendel's law of independent assortment at a lower level. What makes the example especially interesting is that such a description is available not just in principle, but in practice. We already know many of the molecular mechanisms underlying the process of meiosis. The crucial question is: does this knowledge enhance the original cytological explanation? Kitcher recognizes that some additional details will deepen our understanding. For instance, molecular descriptions of the formation of the spindle and the migration of the chromosomes at the poles of the spindle, just before meiotic division, reveal how the chromosomes are not selectively oriented toward these poles. Still, this is roughly where Kitcher draws the line. According to an informal principle of diminishing returns, we soon reach a threshold where providing further molecular details contributes nothing to the explanation.

This point deserves to be clarified. In general, richer biochemical descriptions obviously do contribute to our biological knowledge. Borrowing Kitcher's terminology, molecular biology constitutes an "explanatory extension" of classical genetics. Yet, most molecular depictions of genetic processes do not enhance the explanations offered by higher-level biological theories. The reasons for this failure of transitivity in explanation is that, as noted by Putnam and Fodor, the ascent or descent of levels often yields explanations that do not generalize well. Disjunctive bunches of properties at lower levels typically do not translate into higher-level natural kinds. Unnecessarily detailed descriptions may end up muddying the waters.[13]

The connection between Kitcher's argument and Putnam and Fodor's original insights should now be evident. Atomic properties, it is argued, add nothing to the geometrical square-peg–round-hole explanation. If anything, they hinder the explanatory power of higher-level descriptions. For analogous reasons, biochemical details have no place in higher-level biological theories. At the same time, Kitcher's examples went much further than its predecessors, effectively revealing how the problems for reductionism did not just apply in the abstract, as with Putnam's toy example, or to relatively young and underdeveloped fields, as with Fodor's discussion of the psychoneural interface. The strength of Kitcher's "1953" argument was tackling directly one of the best-case scenarios for reductionism, in one of the most established and faster-growing fields of science. If reductionism could not take flight in genetics, where we already have a clear and detailed picture of the underlying molecular mechanisms, what prospects could there be for disciplines like psychology, sociology, or economics?

In sum, problems with the classical model of reduction spurred an antireductionist consensus according to which the special sciences cannot—and should not—be progressively reduced until we hit the bedrock of fundamental physics. We lack the bridge laws for achieving many such reductions in practice. And, even they were feasible in principle, these derivations would not increase the explanatory power of macro-theories. This *pars destruens* was supplemented with a *pars construens*, which can be characterized, in

---

[13] As Strevens (2016, pp. 156–57) aptly summarizes it in a recent discussion, "Kitcher appears to have the following view: moving down the levels of potential explanation from ecology to physiology to cytology to chemistry to fundamental physics, there is some point beyond which further unpacking of mechanisms becomes entirely irrelevant. [. . .] In '1953' Kitcher accounts for failures of transitivity by proposing that, when transitivity falls through it is because categories essential to explaining high-level phenomena cannot be ascribed explanatory relevance by lower-level explanations due to their not constituting natural kinds from the lower-level point of view."

general, as the explanatory autonomy of the higher levels. The main thesis, pioneered by Putnam and Fodor with an eye to the philosophy of mind, made some form of non-reductive physicalism the default position. These original insights were developed and applied to various other branches of science, such as biology, psychology, and the social sciences.[14]

Still, some would not budge. For one, despite the clear shortcomings outlined earlier, many continued to find the overall reductionist stance convincing. Furthermore, as we shall shortly see, antireductionism is confronted by controversial implications, just like its reductive counterpart. For these and related reasons, the last few decades have witnessed the forging of a "reductionist anti-consensus" across philosophical discussions of the sciences. This new wave of reductionism proposes a refined framework. Moving away from the original logico-positivist model, revamped epistemic reductionism contends that explanations are always improved by shifting down to lower levels. It is time to focus on this theoretical comeback.

§2.5. The Reductionists Return

As discussed in the previous section, antireductionism emerged from problems confronting the classic model of reduction. However, antireductionism is not devoid of difficulties of its own. First, it is often presented as a negative stance, leaving us without a fully developed methodological alternative. Even its main positive core tenet—namely, autonomy—is typically introduced as the in principle impossibility of reduction.[15] Despite strenuous attempts at clarification, concepts such as "explanatory autonomy" and "explanatory extension" remain just as elusive as their reductive counterparts.

Second, the most popular and influential strand of antireductionism—non-reductive physicalism—appears dangerously unstable. The instability, the objection runs, arises because non-reductive physicalism is either not really physicalism, or it is not truly non-reductive.[16] Allow me to elaborate. As a form of physicalism, non-reductive physicalism is committed to the supervenience of all scientific events on physical properties. This is to say that a change in any event described by the special sciences requires some

[14] See Hull (1974); Garfinkel (1981); Kitcher (1984, 1999); Fodor (1999); Kincaid (1990).
[15] For a recent discussion of these issues, see Nagel (2012).
[16] This objection is presented and discussed in detail by Bickle (1998, 2003).

corresponding change in the underlying micro-structure. Yet, *qua* anti-reductionism, non-reductive physicalism eschews the type-identification of properties across levels. These individually plausible assumptions seem jointly in tension. Non-reductive physicalism is a form of property dualism, as it posits non-physical properties. Property dualism comes with a dualist ontology which grounds the claim that even mature physics could not exhaustively explain higher-level events. Thus, critics conclude, non-reductive physicalism becomes a form of dualism, which makes it incompatible with bona fide physicalism.

A third challenge arises from the need to reconcile the success of interdisciplinary endeavors with the autonomy of the special sciences.[17] In one of his early contributions, Fodor noted that "the development of science has witnessed the proliferation of specialized disciplines at least as often as it has witnessed their reduction to physics, so the widespread enthusiasm for reduction can hardly be a mere induction over its past successes" (1974, p. 97). Over four decades later, Fodor's assessment remains accurate. Indeed, it has been reinforced. Rather than being progressively reduced to physics, the special sciences have spawned, developing into a number of burgeoning subfields. The rise of interdisciplinarity is a double-edged sword for antireductionism. For, if the special sciences are "autonomous," as Fodor and many fellow antireductionists maintain, then what explains the constant proliferation of fields such as neurolinguistics, moral psychology, and neuroeconomics, which bring together and integrate independent disciplines?

While all these points deserve discussion, this section focuses on a different objection to antireductionism, more central to our present concerns. The problem challenges the conceptual core of antireductionism, namely, the argument for autonomy. Should we simply accept Putnam and Kitcher's contention that micro-details do not invariably deepen macro-explanations?

Let us begin by clarifying the main issue at stake. As just noted, all properties investigated in the special sciences supervene on physics. Few scholars, including the most fervent supporters of autonomy, would deny this basic materialist tenet.[18] Thus, the crux of the debate over reductionism is independent of physicalism or materialism. These are typically taken for granted, at least in contemporary philosophy of science. The crucial question is the

---

[17] For a more detailed discussion of this point, see Nathan and Del Pinal (2016).

[18] Fodor (1974) calls this thesis "token physicalism," or "the generality of physics," and accepts it as unproblematic. Hütteman and Love (2016) dub it "metaphysical reductionism," to distinguish it from the "epistemological" reductionism presently under discussion.

one first raised in section 1.2 of Chapter 1: in principle, can physics explain everything? Is it true that every fact explained by a special science can be explained just as well—indeed, better, in greater depth—by physics?

Antireductionism offers a negative answer. As Putnam's square-peg–round-hole example purports to show, size, shape, rigidity, and other macro-states supervene on the micro-structure: position, velocity, and other atomic properties. And, yet, the macro-level is superior, from an epistemic stand-point. The physical depiction is less perspicuous than the geometrical one because it cites inessential facts and obscures relevant features. Other influential antireductionism arguments, such as Kitcher's, build on this fundamental insight: the explanatory autonomy of higher-level descriptions.

Not everyone, however, was convinced. One response insists that detailed physical stories do, in fact, provide enhanced illuminations of macro-systems. We all acknowledge that biochemistry explains why taking morphine alleviates pain. Why then would atomic structure not explain the basic geometrical structure of the square-peg–round-hole system? The problem with this line of reasoning is that it quickly turns into a slippery slope. Once we apply it across the board, it turns out that the special sciences seemingly have no explanatory problems of their own. In practice, our limited cognitive capacities may prevent human beings from being able to compute the micro-explanation or from employing it. Yet, *in principle*, physics always provides a deeper account of higher-level events. From this stand-point, the autonomy of the special sciences is a mere pragmatic byproduct of our ignorance or lack of computing power. Many reductionists suggest that the theoretical disposability of higher levels does not undermine the intrinsic value of biology, neuropsychology, or economics, vis-à-vis the omnipotence of physics.[19] Antireductionists typically beg to disagree.

A stronger objection turns Putnam's argument on its head, dismissing allegedly "autonomous" geometrical explanations as either incomplete or straight up false.[20] The effectiveness of geometrical depictions, the story goes, presupposes a host of physical information regarding, say, the rigidity of materials and their behavior in the conditions explicitly presented—or tac-itly assumed—in the description of the system. As these depictions are progressively completed, filling in gaps, the crucial relevance of the fundamental physical states and corresponding laws will become more evident.

[19] For a response along these lines, see Sober (1999, 2000).
[20] This rejoinder to Putnam's argument is found in Rosenberg (2006, p. 35, fn. 3).

Related worries have been raised against Kitcher's arguments developed in his article titled "1953 and All That." As discussed in section 2.4, Kitcher questioned the possibility of reducing classical Mendelian genetics to molecular biology. Succinctly put, does molecular theory really not explain the classical theory of genetics? Are the shallower explanations of Mendelian genetics objectively preferable and more powerful than the detailed accounts provided by current biochemistry? Molecular biology has greatly improved our understanding of gene replication, expression, mutation, and recombination. Does this not make the antireductionist perspective outdated? Why such pessimism? Why not expect molecular biology to continue on its path toward explaining genetics in accordance with the spirit, if not the form, of post-positivist reduction?[21]

These debates were hardly confined to biology. A similar interplay or arguments and counterarguments can be seen, around the same time, in philosophical discussions of mind, psychology, and neuroscience. In the first half of the twentieth century, Descartes's mind-body problem was reformulated in terms of reduction. Can mental states be reduced to brain states? Initial attempts by U. T. Place and J. J. C. Smart to spell out a version of psycho-neural reduction in accordance to the post-positivist model were thwarted by the same problems discussed earlier in section 2.3: multiple-realizability, lack of laws, and the like. This led to the formulation of various antireductionist theories of mind, such as functionalism and the token-identity theory. But not all scholars viewed the endorsement of autonomy as the appropriate response. Identifying the problem of early type-identity approaches with their crudeness, this neo-reductive wave in the philosophy of mind produced several alternative approaches, such as eliminative materialism, functionalization, and various revised versions of the identity theory.[22]

---

[21] This is Ken Waters's (1990) response to Kitcher's "explanatory incompleteness objection." Waters also responds to another class of antireductionist arguments, which attempt to establish an unbridgeable conceptual gap between these areas of biology because of subtle differences in the meaning of parallel terms (Hull 1974; Rosenberg 1985). The main idea underlying this "unconnectability objection" is that, while both classical geneticists and molecular geneticists talk about "genes," the term is used very differently across these theories. Mendelian genes are identified through their phenotypes. And the relation between molecular genes and phenotypes is exceedingly complex, frustrating any systematic attempt to connect the two concepts along Nagel's lines. Waters responds by arguing that this unconnectability objection presupposes an oversimplified conception of Mendelian genes. Once the original theory is correctly understood, he claims, "The Mendelian gene can be specified in molecular biology as a relatively short segment of DNA that functions as a biochemical unit. [...] I conclude that the antireductionist thesis that there is some unbridgeable conceptual gap lurking between [Classical Mendelian Genetics] and its molecular interpretation is wrong" (1990, p. 130).

[22] Eliminative materialism was jointly developed by Patricia and Paul Churchland (1979, 1986). On this view, mental states are theoretical entities posited by "folk psychology." Commonsensical as it seems, folk psychology, they argue, is a flawed theory of mind and is thus not a credible candidate

Time to take stock. This chapter began with a discussion of classical reductionism and its well-known shortcomings. Next, we presented the gradual emergence of an antireductionist consensus. While antireductionism remains an influential player in the game, it is no longer hegemonic, as its foundations have been challenged by a resurgence of the reductionist perspective. This revitalized reductionism acknowledges the limits of the classical model. It contends, however, that this was never the real issue at stake. Reductionism, its new-wave defenders claim, should not be abandoned. It should be reconfigured, from a logical relation between theories to an epistemic tenet regarding scientific explanation. This shift is evident in the writing of many influential contemporary authors.[23] But the main questions remain largely unsettled. Do micro-explanations always deepen the explanatory power of macro-depictions? Or are the higher levels epistemically autonomous from the lower levels, as suggested by Putnam and his followers? Can this long-standing debate be settled, once and for all?

§2.6. Scylla or Charybdis?

So, who wins? The reductionist or the antireductionist? Are we going to get crushed by the rock or drowned by the whirlpool? The answer hinges on whether we can—and should—describe every scientific event at more fundamental levels, and whether these micro-depictions invariably enhance

for integration. Rather, it should be eliminated and replaced by a developed neuroscience, which will be more predictive, more explanatory, and better connected with extant scientific research. Eliminative materialism is, at the most general level of description, a detailed and provocative attempt to apply a revised account of classical reduction to the traditional mind-body problem. To be sure, eliminativists prefer to talk about "elimination," rather than "reduction." Yet, the former can be treated as a limiting case of the latter. Other notable reductionist strategies include Lewis's (1972) and Kim's (2005) sophisticated versions of identity theory and Bickle's (1998, 2003) "ruthlessly reductive" account.

[23] For example, Rosenberg (2006, p. 12) characterizes biological reductionism as the tenet that "there is a full and complete explanation of every biological fact, state, event, process, trend, or generalization, and [. . .] this explanation will cite only the interaction of macromolecules to provide this explanation. This is the reductionism that a variety of biologists and their sympathizers who identify themselves as antireductionists need to refute." Or, again, "the impossibility of postpositivist reduction reveals the irrelevance of this account of reduction to contemporary biology, not the impossibility of its reduction to physical science. Thus, the debate between reductionists and antireductionists must be completely reconfigured" (Rosenberg 2006, p. 22). Similarly, in a recent article addressing Kitcher's arguments, Strevens (2016) suggests that the in-practice autonomy of the special sciences can be made compatible with reductionism. This requires understanding the high-level sciences' systematic explanatory disregard of lower-level details of implementation as practically, as opposed to intellectually, motivated. For similar stances, see Schaffner (1967, 1993, 2006) and Hooker (1981).

explanatory power. Once again, clear-cut answers are still wanting, and not for lack of trying. How come? The reason, I maintain, has largely to do with the two parties talking past each other. Allow me to elaborate.

A popular litmus test to locate one's position on the reductionism-antireductionism spectrum is to make an informed prediction concerning the future state of science. This is because no one really wants to argue that *current* physics is in a position to replace biology, psychology, economics, or any other branch of science. We lack the deep understanding of subatomic systems and the computing power required to approximate the perfect vision of a Laplacian Demon. Still, reductionism claims, in principle, it would be possible to provide a host of physical depictions that could not merely replace extant higher-level accounts but enhance their explanatory power. Antireductionists beg to disagree. Even if we were in a position to translate biological, psychological, or economic generalizations in molecular terms, we should not do so. If anything, this would muddy the waters.

This is how the debate was first introduced in section 1.2 of Chapter 1. Intuitive and thought provoking as they are, these questions have a shortcoming: they are untestable. The reason is simple. How do we assess, at our present time, the fate of future science, that is, its development in the long run? To adjudicate between these competing stances, we need to look at whether current science better conforms to the standards of reductionism or antireductionism. When we try to do so, the matter becomes less substantive than most participants like to admit. To corroborate this claim, I focus, in turn, on how the dispute has unraveled in two fields: biology and psychology.

Can all biological events be subsumed under a physical explanation? Unfortunately, this question cannot yet be answered, and this is unlikely to change anytime soon. Sure, we are starting to see some fruitful interdisciplinary overlap between physics and biology. Still, the truth is that we are too far from integrating these two disciplines to warrant an informed response, one way or the other. For this reason, the fate of reductionism in biology hinges on a more modest thesis, namely, *molecular reductionism*.[24] According to this tenet, there is an explanation of every biological fact that mentions only biochemical properties of molecules and their interactions.

Whereas the general reductionist query—can physics explain everything?—pertains to the realm of science fiction, molecular reductionism may be assessable in practice, not merely in principle. As noted in section 2.5,

---

[24] This thesis is spelled out in detail in Sarkar (1998) and Rosenberg (2006).

several functional biological concepts have already been characterized from a molecular perspective: gene replication, expression, mutation, recombination, and so forth. Now, to be sure, even if the truth or falsity of molecular reductionism could be successfully established beyond reasonable doubt, we would still be far from a vindication or rejection of the *tout court* reductionism envisioned by logical positivism. Still, failure (success) in a specific field would provide evidence for (against) the overarching thesis.

So, is molecular reductionism a plausible assumption? Is there an explanation of every biological fact that appeals only to a purely structural language, eschewing functional concepts and any other higher-level term?

The trouble with this question is that a clear-cut answer presupposes a rigorous definition of "molecular concept" and "purely molecular language," which we lack. These expressions turn out to be dangerously vague. Precise characterizations have never been articulated, and it is unlikely that they will ever be found. It is important to stress that the problem does not lie with the explanations themselves. My claim is not that explanations in molecular biology are imprecise or otherwise inaccurate. Quite the contrary. What is unclear is whether the explanations in question should be classified as "structural" or "purely molecular" and, therefore, whether they adhere to the reductionist paradigm or to its antireductionist counterpart.

Some illustrations should help drive the message home.[25] Consider the typical explanation of drepanocytosis, a disease commonly known as sickle-cell anemia. Its potentially devastating effects are triggered by a point mutation, a single-base substitution in the chain of nucleotides of an individual gene, the *hemoglobin* gene. In short, we have a genetic mutation that produces a physiological dysfunction, due to a structural modification of cells. This looks like a textbook case of reduction. If this story does not conform to the molecular reductionist paradigm, I am not sure what will.

Upon further scrutiny, things turn out to be much more complicated than they initially appear. Is drepanocytosis a successful instance of molecular reduction? The key is whether only structural biochemical concepts figure in the explanans. Is this the case? Standard textbook explanations of sickle-cell anemia appeal to various dispositional properties, such as the flexibility of erythrocytes and the elasticity of capillaries. Similarly, terms like "transcription factor" involve an implicit appeal to function, that is, the regulatory role of these proteins in the transcription of genes. Translating all dispositional

[25] For a more detailed discussion of these examples, see Nathan (2012).

and functional notions in structural terms would be anywhere from frustratingly hard to impossible. Are all these properties part of a molecular language? Or do such appeals to non-structural concepts reveal that even the best-case scenarios for the reductionist are doomed to fail? What seemed a cut-and-dried empirical question now looks like a matter of stipulation: what counts as a "purely structural" or "molecular" language?

Things are about to get worse. Once we move away from relatively straightforward genetic explanations—arguably, the clearest examples of molecular-reductionism in biology—the question of whether the concepts used are "strictly biochemical" becomes, if anything, more quixotic. Think about reaction-diffusion models, which depict chemical reactions between types of molecules as they spread throughout a system. Are the mathematical models describing the geometry of the diffusion-space part of the language of molecular biology? What about textbook explanations of the segmentation of *Drosophila*, where difference-makers simultaneously occur both at the molecular and geometrical levels?

The examples themselves might seem complicated, but the underlying point is simple. A cursory glance at current biological research reveals how explanations help themselves to a variety of concepts: structural properties of genes and gene products, functional-dispositional features of reactants, spatial-geometrical parameters of systems, and diffusion processes. Are all of these part of a "fully molecular language"? How do we even begin answering this question objectively, that is, without appealing to stipulation?

What conclusions should be drawn? The question presently under consideration is: does biological practice vindicate molecular reductionism? A clear answer, I suggested, is thwarted by the lack of clear definitions of "molecular language" and derivative terms. Reductionists typically maintain that functional concepts, dispositional terms, and whatever else is required to explain biological systems are part of molecular biology and its toolkit. Antireductionists contend that this heterogeneous conceptual toolkit transcends the pure biochemical realm, as the ubiquitous appeal to teleology and function thwart any characterization of a "fully molecular" language.[26] If both parties can agree on the range of explananda and the required explanantia, then what exactly is under dispute? Is it the nature of biological explanation or whether or not to label these explanations "molecular"?

[26] These divergent characterizations of a "fully molecular" language can be found in the work of Rosenberg (2006) and Schaffner (2006), on the reductionist side, and Culp and Kitcher (1989), Kincaid (1990), and Franklin-Hall (2008), from the opposite perspective.

It is important to stress that the issue is independent of biology. Analogous problems affect similar debates across the board. As noted in section 2.5, contemporary philosophy of mind has been dominated by the question: can mental states, broadly construed, be reduced to brain states, broadly construed? Behaviorists, type-identity theorists, and eliminativists answer in the positive. Psychological states, they contend, are reducible to neurological ones. Token-identity theorists, functionalists, and property dualists answer in the negative, arguing that any such reduction is doomed to fail.

If reductionism is so central to the philosophy of mind, one could legitimately ask, where do we currently stand on psycho-neural reduction after almost a century of extensive debate? A glance at the relevant literature reveals that not much success can be boasted. Now, we have surely made significant progress in unveiling the psychological and neural mechanisms underlying both higher and, especially, lower cognition. Have these results tipped the scale with respect to the question of the reduction of mental states to brain states? If so, the news has not yet been broken, as there seems to be no more consensus on this issue than there was in the 1950s.

When confronted with this lack of resolution, many scholars interested in the nature of mental states—scientists and philosophers alike—respond by pointing to the extreme intricacy of the subject matter. The human brain is arguably the most complex system discovered so far in the universe, with billions of cells and an astronomical number of possible connections among them. No wonder that solving these disputes is proving so hard!

I wholeheartedly agree. I am nevertheless skeptical that the complexity of the structures under investigation is responsible for the lack of tangible progress in the philosophy of mind. The main culprit, I maintain, is the alleged contrast between *reductionism* and *antireductionism* themselves.

Can theories, concepts, or laws in psychology be "reduced" to the level of neuroscience? Once again, this question is in need of clarification. What exactly are we asking? If reduction is understood along the lines of the classical model, then the answer is clearly negative. Let us all agree, once and for all, that the principles of psychology cannot be translated into the principles of neuroscience via Nagelian bridge laws, in practice or in theory. Case closed. If the question is interpreted as whether we can and should provide a "functionalized" characterization of mental states then, all of a sudden, an affirmative answer looks more promising.[27] Yet again, if the issue is whether or not all psychological events could and should be explained more thoroughly by describing them in neuroscientific terms, the answer depends on how we choose to characterize psychology and neuroscience. If the "language of neuroscience" is restricted to talking about individual neurons and their additive interactions, this is clearly insufficient to explain psychology, period. Indeed, such a narrowly construed neuroscience would not even be powerful enough to describe neural events, let alone mental ones. As we shall discuss, in greater detail, in Chapter 8, neural systems, as currently understood, are irreducible to individual causes and transcend the localization and decomposition of their elements. In contrast, sufficiently rich characterizations of the appropriate vocabulary—including functional, behavioral, and cognitive concepts—will make much of psychology describable in neural terms, not only in principle, but in practice as well. And such a reduction may well be fruitful and insightful.

It is important to emphasize, once again, that the vagueness affecting current debates on psycho-neural reductionism is not grounded in factual ignorance. Sure, cognitive neuroscience is, relatively speaking, a young discipline and much still needs to be learned. Nevertheless, a look at more established fields suggests that lack of knowledge cannot be the main issue at stake. Consider, once again, the situation in biology. Biologists are often able to successfully pinpoint the implementation of functional structures at the molecular level. We already know quite a bit about, say, how phylogenetic adaptations develop at the ontogenetic level. Many complex conditions can be causally explained by identifying their genetic difference-makers and the subsequent cascade of processes. Prima facie, this might suggest that the case for or against molecular reductionism has been settled, or is getting close to being settled, in the life sciences. Yet, as we saw earlier, this is not the case. The debate over molecular reductionism is as open as ever. For this reason, we should not expect advances and discoveries concerning where and how cognitive functions are computed in the brain to solve the question of psycho-neural reduction. Philosophers of neuropsychology should learn the hard lesson from their colleagues in biology.

Before exploring alternatives, it is time to tie up some loose ends.

[27] From a functional perspective, to "reduce" a higher-level theory to a lower-level one is to provide an analysis of the macro-properties in terms of inputs, outputs, and the realization of their causal role at the more fundamental level (Kim 2005).

§2.7. Why Not Have the Cake and Eat It Too?

To wrap things up, let us return to the question that has fueled our discussion. Who wins, the reductionist or the antireductionist? Scylla or Charybdis? The bulk of this chapter laid out several ways of presenting the two stances, and various arguments for and against them. Section 2.6 argued that the question of whether macro-explanations can be enhanced by rephrasing them at the micro-level has no clear, definitive answer. The issue depends more on terminology or linguistic preference than substantive disagreement. Paraphrasing one of Wittgenstein's aphorisms, reductionism and antireductionism are matters of expression, not facts of the world.[28] Are there more substantive disagreements to be found? I believe there are.

Both reductionism and antireductionism center on explanation. This strikes me as correct. Still, the interface between higher and lower levels, implicitly adopted by both parties, is overly restrictive. Reductionists suggest that micro-explanations invariably advance macro-depictions. Antireductionists respond that some higher levels exhibit an epistemic autonomy of sorts, in the sense that they are perfectly explanatory without the addition of further details that would only muddy the waters. Autonomy and reduction are typically assumed to be mutually exclusive. Indeed, autonomy is routinely defined as the rejection of reduction and, vice versa, reduction is characterized as the rejection of autonomy. I want to argue that these two tenets may actually be reconciled. What does this entail?

Compared to geometrical accounts, physical descriptions provide more comprehensive understanding of why square pegs do not go through round holes whose diameter is approximately the length of their cross section. Still, the geometrical explanation delivers the goods in a more succinct fashion, omitting unnecessary detail. How is this possible? Do these observations not flatly contradict each other? Contrary to conventional philosophical wisdom, I shall answer these questions in the negative. By focusing not on what our models state, but on what they leave out, it will become clear that autonomy and reduction are really two sides of the same coin.

---

[28] Here is Wittgenstein's original quote: "The fallacy we want to avoid is this: when we reject some form of symbolism, we are inclined to look at it as though we had rejected a proposition as false. It is wrong to treat the rejection of a unit of measure as though it were rejection of the proposition. "The chair is three feet rather than two". This confusion pervades all of philosophy. It is the same confusion that considers a philosophical problem as though such a problem concerned a fact of the world instead of a matter of expression" (1979, p. 69).

The reconciliation of reduction and autonomy will provide a more systematic explanation of the interplay between productive ignorance and knowledge. The reason is that neither reductionism nor antireductionism, by itself, captures one side of the equation, but remains moot on the rest of the story. Reductionism, it should now be clear, captures well the image of science as a wall progressively constructed in a brick-by-brick fashion. Science advances by breaking down complex macro-systems into smaller and smaller building blocks, until we reach bedrock. This approach, unfortunately, seems to leave little to no room for productive ignorance, as replacing a vague characterization with more detailed descriptions invariably constitutes a step forward. The notion of autonomy underlying antireductionism, in contrast, legitimizes the role of ignorance in science, by stressing how the addition of micro-information does not add anything to the explanation. If anything, it muddies the waters. Yet, as discussed, the claim that higher-level explanations can be objectively superior to lower-level ones remains unpersuasive. In short, a plausible account of productive ignorance needs to combine these two perspectives, not pitch one against the other. Seeing how this can be done, however, will require some work.

In conclusion, it is time to sail away from the Scylla of Reductionism and the Charybdis of Antireductionism. I propose to do so by focusing on Firestein's question introduced in Chapter 1. How does science turn ignorance into knowledge? Our first step will be to illustrate my general thesis with some examples from the history of science. We will come back to the rock shoal and the whirlpool in Chapter 10, with much more under our belts. By then, paraphrasing the lyrics of Judas Priest, you'll be heading out to the highway, ready to weather every storm that's coming atcha.

3

Lessons from the History of Science

I know, for I told me so,
And I'm sure each of you quite agrees:
The more it stays the same,
The less it changes!

—Spinal Tap, "The Majesty of Rock"

§3.1. Introduction

Chapter 1 started to draw attention to the place of ignorance in science. Can ignorance play a genuinely productive role? How? Why is this positive contribution systematically overlooked, to the point of mischaracterizing the scientific enterprise in textbook presentations? Chapter 2 identified the main culprit as the reductionism vs. antireductionism rhetoric, which underlies much contemporary philosophy of science. First, the disagreement is largely an illusion. Antithetical as they may seem, sophisticated versions of these stances overlap, making the opposition less substantive and more semantic than is typically assumed. In addition, neither strand adequately captures the interplay between knowledge and ignorance. The cumulative reductionist model simply leaves no space for ignorance. Antireductionism, in contrast, accommodates ignorance, but fails to show how it may play a productive role in the advancement of science.

My alternative proposal revolves around an epistemic strategy that is oft mentioned, but seldom analyzed explicitly: a form of explanation called *black-boxing*. A systematic analysis of black boxes will be postponed until Chapters 4 and 5. Before getting there, I need to convince readers that these constructs are as prominent as I claim them to be. To this effect, this chapter presents four episodes in the history of science where black-boxing takes center stage. These case studies show that our strategy is not a single, monolithic practice merely intended to mask lack of knowledge. It is a flexible and multifaceted process that accomplishes a broad range of tasks.

Black Boxes. Marco J. Nathan, Oxford University Press. © Oxford University Press 2021.
DOI: 10.1093/oso/9780190095482.003.0003

§3.2. Darwin's Quest for Inheritance and Variation

When one speaks of Darwinism today, what is usually meant is *evolution by natural selection.*[1] Natural selection arguably was Darwin's most daring and novel idea, underlying his ambitious account of apparent harmony and adaptation in the organic world. What made the proposal so revolutionary?

Darwin's theory of evolution by natural selection provides a powerful framework for explaining patterns of distribution of species and traits across the globe. The main conceptual ingredients are introduced in the opening four chapters of *On the Origin of Species* (1859). Here is a brief overview.[2]

The first *principle of variation* states the obvious: within every species, there is always variation. The second *principle of competition* notes that because of the tendency of populations to grow exponentially, at almost any stage in the history of a species, more organisms are born than can survive and reproduce. This is essentially the "struggle for existence" theorized by Malthus and other "classical economists" of the time. The third *principle of variation in fitness* notes that some of the differences between organisms, captured in the first principle of variation, will affect the organism's capacity to survive and reproduce. By combining these ideas with his fourth *principle of inheritance*, according to which some of this variation is heritable, Darwin was able to derive the idea of differential survival. From these four simple premises follows a fifth statement, the celebrated *principle of natural selection*: heritable variations that improve the ability to survive and reproduce become more prevalent over successive generations.[3]

Notably, none of these tenets—natural selection included—was original. All five propositions would have been accepted by most mid-nineteenth-century intellectuals. Darwin's insight thus does not consist in his formulation of evolution by natural selection. As Kitcher argues, the breakthrough

---

[1] Following conventional wisdom, I refer to "evolution by natural selection" in the singular, treating it as an individual, monolithic, and cohesive theory. Yet, strictly speaking, the "theory" of evolution encompasses a package of independent theses. These include the constant presence of reproductive surplus, the continuous production of individual differences, the heritability of traits, sexual selection, and a few other tenets (Mayr 1991).

[2] The following reconstruction of Darwin's tenets is inspired by Kitcher (1985, 1993).

[3] Darwin has a simpler conception of natural selection, compared to contemporary evolutionists (Mayr 1991). For Darwin, selection is essentially a one-step process, which involves a steady production of individuals, generation after generation. Some of these are bound to be "superior," in virtue of having some reproductive advantage over competitors. Modern mainstream evolutionary theory, in contrast, depicts natural selection as a two-tiered process. The first step consists in the generation of genetically distinct individuals, the production of variation. The second step is the actual selection process, which determines the survival and reproductive success of these organisms themselves.

rather lies in Darwin's pioneering understanding of the far-reaching implications of evolution, which is capable of breaching species boundaries. Essentially, Darwin appreciated how the simple phenomenon of differential survival, over many generations, becomes the engine for the evolution of species.

A comprehensive overview and discussion of evolution by natural selection lies well beyond the scope of this work. The important historical point, for present purposes, is straightforward. The availability of an unlimited supply of variation, especially heritable variation in fitness, is a cornerstone of Darwin's own formulation of evolutionary theory. Natural selection requires the existence of intra-species variation, that some of this variation affects the capacity of organisms to survive and reproduce, and that some of these differences be passed on from one generation to the next. Indeed, once natural selection is assumed as the chief engine of evolutionary change, the only missing piece in Darwin's puzzle is a reliable set of processes to generate and spread the required heritable variability within and across populations.

So, what is the source of all this heritable variation in fitness? This question preoccupied Darwin for his entire career. Merely recognizing the abundance of differences within the biological realm did not satisfy the great naturalist, who refused to accept spontaneous variation or the manifestation of the imperfection of the organic world as plausible accounts. Darwin rightly demanded more. He realized how the fate of his theory depended on finding a plausible causal explanation for the mechanisms of variation and inheritance. While Darwin admittedly privileged the former over the latter,[4] I momentarily set variation aside and focus mainly on inheritance.

Until late in the nineteenth century, it was widely believed that both environmental forces and usage could affect the heritable qualities of traits. The underlying intuition—frequently albeit misleadingly dubbed "inheritance of acquired characters"[5]—is that the material basis of inheritance, what has now

---

[4] Darwin considered inheritance and its laws as a less-immediate concern than variation and its causes, which troubled him from the initial stages of his evolutionary thinking. Two aspects of variation were especially problematic (Mayr 1982). First, while Darwin was clearly thinking about the range of variation for domesticated species, he never draws a clear distinction between interpopulational and intrapopulational variation, that is, between individual and geographical varieties. Second, although Darwin acknowledges the existence of discontinuous variation, he stresses the prevalence and biological significance of continuous variation. It was genetics which eventually showed that there is no fundamental difference between continuous and discontinuous variation.

[5] This label is imprecise because such a view typically included the modifiability of genetic material by climatic and other environmental conditions ("Geoffrism") or by nutrition directly, without the intermediary role of phenotypic traits (Mayr 1982, p. 687).

been identified as genetic material, broadly construed, is pliable or "soft." Soft inheritance was so commonly taken for granted that hardly anyone even bothered to actually try to figure out its mechanistic underpinnings before the 1850s. Justifying the existence of soft inheritance, however, became of paramount importance with the emergence of the evolutionary framework, as substantial questions now hinged on it. Is evolution fueled by the inheritance of acquired characters, as proclaimed by the French evolutionist Lamarck and his followers? Or are there different processes at play?

As previously noted, Darwin was explicitly committed to most individual variation being heritable. At the same time, contrary to popular belief, throughout his life he accepted both soft *and* hard inheritance, changing opinion only on the relative importance of the two. Even mature works such as the *Origin* occasionally cite evidence in favor of soft inheritance.[6] Incidentally, in this respect, Darwin's own account of the material basis of evolution is closer to Lamarck's than many contemporary readers realize.

What was Darwin's view of the basic laws and mechanisms of inheritance? In the *Origin*, he candidly confesses his ignorance on the matter:

> The laws governing inheritance are quite unknown; no one can say why a peculiarity in different individuals of the same species, or in individuals of different species, is sometimes inherited and sometimes not so; why the child often reverts in certain characters to its grandfather or grandmother or other more remote ancestor; why a peculiarity is often transmitted from one sex to both sexes, or to one sex alone, more commonly but not exclusively to the like sex. (1859, p. 13)

The tone, however, becomes less cautious in later works. In Chapter 27 of *Variations of Animals and Plants under Domestication*, Darwin sketches his own theory of inheritance. This is a "provisional hypothesis of pangenesis," geared toward explaining the plethora of variation within and across species. Briefly, Darwin surmises that the transmission of heritable qualities, as well as the unfolding of development, is caused by individually different particles, small to the point of being invisible, which he calls "gemmules." In order to accommodate the possibility of organisms inheriting acquired characters, either from the environment or through use and disuse, the view

---

[6] As stressed by Mayr (1982), Darwin discusses three potential sources of soft variation: changes in the environment that induce increased variability via the reproductive system, unmediated influences of the environment, and the effect of use and disuse.

is supplemented with a "transmission hypothesis." At any stage of the life cycle, he suggests, cells may throw off gemmules and, when supplied with nutrients in sufficient quantity, they multiply by self-division.

To be sure, Darwin was rather reticent about his own account of pangenesis.[7] This caution turned out to be well-placed. His theory was first rendered obsolete and subsequently disproven by the German evolutionist August Weismann, who identified the nucleus as the vehicle of inheritance. Roughly fifteen years after the formulation of Darwin's "provisional hypothesis of pangenesis," Weismann successfully unraveled the complex structure of the chromatin inside the nucleus, conclusively rejecting the very possibility of soft inheritance. Unfortunately, Darwin and his British colleagues were unaware of the spectacular advancements in cytology in *Mitteleuropa* at this time. Consequently, Darwin remained quite uncertain throughout his life about the nature of continuous variation, which remained controversial until genetics finally settled the issue after 1910.

In sum, abundance of variation, especially heritable variation in fitness, was a cornerstone of Darwin's theory of evolution by natural selection. The nature and source of inheritance and variation concerned the great naturalist throughout his long and productive career, leading him to provide his own speculative hypotheses. One can hardly fault Darwin for what he admittedly did not know or failed to grasp. Lacking any deep understanding of genetics, whose rudiments had to await the rediscovery of Mendel's work at the turn of the twentieth century, Darwin's explanations in *Origin* made no reference to mechanisms of inheritance or variation. Furthermore, many of his subsequent proposals strikingly missed the mark. Pangenesis and, in particular, the transmission hypothesis did not withstand serious scrutiny.

These well-known observations raise an interesting quandary. Given that, by Darwin's own light, inheritance plays a crucial role in ontogeny and, indirectly, in phylogeny, should one not expect his evolutionary explanations to be dreadfully inadequate? Somewhat surprisingly, the answer is negative. Glossing over *Origin*, many contemporary readers are struck by the variety and detail of examples collected therein, together with the depth of Darwin's insightful explanations. Topics range from traits changing function in the course of evolution (what we now call "exaptations"), to comparative analyses of covariation occurring in distant albeit structurally similar environments.

---

[7] According to Mayr (1982, p. 694), he was especially wary about the transmission hypothesis, which he referred to as a "mad dream" that nonetheless "contains a great truth."

One also finds rudimentary, and yet remarkably plausible, hypotheses regarding the formation of complex organs that require myriad generations to evolve but could not obviously do so in a piecemeal fashion. What use are a quarter of a wing or a not fully functional eye? *Origin*'s answer turned out to be on the right track.

In short, Darwin was aware of the role of variation and inheritance in his theory of evolution and realized how much information he lacked. He also later integrated his views with speculative and, for the most part, incorrect assumptions. How is it possible for his explanations to be so successful?

This puzzling aspect of Darwin's work has not gone unnoticed among scientists, philosophers, historians, and other scholars interested in biology. A clear statement of the main issue, together with a sketch of a solution, can be found in Mayr's seminal excursus into the growth of biological thought:

> The most fascinating aspect of Darwin's confusions and misconceptions concerning variation is that they did not prevent him from promoting a perfectly valid, indeed a brilliant, theory of evolution. Only two aspects of variation were important to Darwin: (1) that it was at all times abundantly available, and (2) that it was reasonably hard. Instead of wasting his time and energy on problems that were insoluble at his time, Darwin treated variation in much of his work as a "black box." It was forever present and could be used in the theory of natural selection. But the investigation of the contents of the box, that is, the causes of the variation, occupied Darwin only occasionally and with little success [. . .] Fortunately for the solution of the major problems with which Darwin was concerned [. . .] a study of the contents of the box was unnecessary. It could be postponed until more auspicious times. (Mayr 1982, p. 682)

Mayr can be paraphrased as claiming that the two projects of identifying the mechanisms of transmission and variation and of spelling out the processes of evolution are not inextricably tied to each other. One can acknowledge the existence of hereditary variation and its fundamental biological role while effectively setting its nature and structure aside. This seems plausible. Yet, Mayr's observation raises a further philosophical question. *How* is it possible to segregate the mechanisms of inheritance—a notion which lies at the very heart of the phenomenon we are explaining: evolution by natural selection—without compromising the integrity of the explanations, making them partial, superficial, or otherwise inadequate?

Mayr's insight already contains the key to answering this question. Darwin treats inheritance and variation as black boxes. The variability and heritability of certain traits is undeniable. Darwin was clear and explicit about this. Nonetheless, he deliberately set aside and postponed puzzles concerning their nature and structure until more auspicious times. And, when he did try to answer these issues, in later publications, his wrongheaded speculations left his evolutionary analyses unscathed.

Taking stock, Darwin stressed the role of variation and heritability in his theory. Still, he "black-boxed" the underlying mechanisms. He acknowledged their existence and significance but took them for granted. Now, surely Darwin would have loved to know more about these processes. Nevertheless, the concepts of inheritance and transmission, as they figure in *Origin*, did not require any micro-explanation to be effectively employed within evolutionary theory. All this is well known. But how exactly does it work?

The aim of this book is to spell out the foundations and implications of this black-boxing strategy. What is a black box? How do we create one? Under what conditions is this epistemic construct legitimate, successful, or justified? Are there better or worse ways of isolating a phenomenon? Before delving into the nuts and bolts of our explanatory strategy, I would like to introduce a few more examples. These fascinating case studies emphasize how pervasive black-boxing is across disciplines. In addition, they reveal a multifarious array of reasons underlying the deliberate decision to leave certain phenomena unspecified, while unveiling others. As we shall see, these motivations are vastly richer, more complex, and more exciting than Mayr's mere need to "postpone and await for more auspicious times."

§3.3. Mendel's Missing Mechanisms: A Tale of Two Syntheses

On the Origin of Species marks a true watershed in the history of biology. It provides a powerful recipe for explaining the evolution of organisms and traits across the globe. Still, Darwin left many questions unanswered. It thus became of paramount importance for researchers following in his wake to obtain a detailed understanding of the mechanisms governing the variation and transmission of traits. Unveiling them was a slow and painstaking achievement. A seminal contribution came from Mendel's studies that, incidentally, provide another paradigmatic illustration of black-boxing.

In his pioneering experiments with pea plants, Mendel investigated the process by which traits are passed on from one generation to the next. His observations were crystallized into three famous principles, here paraphrased in contemporary biological jargon. First, the *law of segregation* states that during gamete formation, the alleles for each gene segregate one another so that each gamete carries only one allele per gene. Second, the *law of independent assortment* (previously encountered in §2.4 of Chapter 2) asserts that genes located on non-homologous chromosomes segregate independently during the formation of gametes. Third, the *law of dominance* distinguishes between "dominant" and "recessive" alleles by stressing that an organism with at least one dominant allele will display its effect over the recessive variant.

A comprehensive discussion of these celebrated results transcends the scope of this work. Our key observation is that Mendel did not provide any description of the mechanisms responsible for the transmission of "factors." In fairness, Mendel was not seeking a mechanistic characterization and, arguably, he did not even have a sophisticated concept of mechanism in mind. In accordance with his training in physics within the Austro-Hungarian school, he was after a mathematical description of laws and their consequences, which could capture and systematize his meticulous data. Furthermore, Mendel predated many new cytological findings, many of which were accomplished in the 1870s and 1880s, years after the publication of his paper.

Still, this raises a question. Given his ignorance of basic cytology, how could Mendel explain the transportation of characters in gametes? Simply put, he did not explain it! As Mayr (1982, p. 716) eloquently puts it:

> He [Mendel] postulated that the characters are represented by "gleichartige [identical] oder differiende [differing] Elemente." He does not specify what these "Elemente" are—who could have done so in 1865?—but considers this concept sufficiently important that he refers to these "Elemente" no less than ten times [. . .] [in] the *Versuche*."

The obvious follow-up is: why is Mendel unanimously considered the father on modern genetics, given that he did not understand the fundamental structure of genes any better than other fellow naturalists of his time?

In response, Mayr stresses two crucial differences between the Bohemian scientist and illustrious colleagues of his, such as Darwin, Galton, and Weismann. First, Mendel discovered a consistent 3:1 ratio in inheritance patterns. This effectively refuted the widespread "multiple particles postulate," according to which numerous identical determinants for a given unit character are present in each cell, and many replicas of a single determinant might be transmitted simultaneously to the germ cell. Second, Mendel realized that these particles exist in sets—genes and their alleles, we would say today—which allowed an explanation of segregation and recombination.

In short, Mendel did have a deeper understanding of variation and heritability, compared to other biologists of his time, Darwin included. Yet, this does not fully capture his achievements. Mendel's knowledge involved more accurate generalizations that constrain the behavior of underlying entities. But, from a mechanistic standpoint, the Bohemian scientist black-boxed variation and inheritance no less than his illustrious English colleague. And, contrary to Darwin, this did not seem to bother Mendel in the least.

The key to Mendel's call to fame lies in the story of how biology opened his black box: the gene. This is a tale that involves more black boxes.

Mendel's groundbreaking work, published in 1866, remained ignored by many naturalists until its "rediscovery" in 1900. Even then, it took decades for genetics to finally prosper as an independent discipline. In the meantime, the life sciences were becoming increasingly fragmented. Researchers in different biological fields approached the study of evolution at dramatically different scales. Paleontologists focused on the macro-picture: the fossil record and the phylogeny of higher taxa. Systematists were largely concerned with the nature of species and the process of speciation. Early geneticists, in turn, studied the variation of traits in a handful of model organisms—most notably various species of the fruit fly *Drosophila*. All these disciplines were disconnected and, at times, even mutually hostile. Colleagues with offices across the hall from each other bickered over who provided the deepest, most fruitful insights into the process of evolution.

Harmony was eventually achieved through the integration of all these evolutionary perspectives, pitched at different levels. This achievement can be broken down to two broad phases. The first stage occurred in the 1920s and 1930s, when population geneticists such as Fisher, Haldane, and Wright reconciled the conflict between "saltationism," the discontinuity of characters of Mendelian alleles, with the continuously varying characters postulated by traditional Darwinism. The second phase, which began in the late 1930s and continued well into the 1970s, witnessed the ecumenical efforts of Dobzhansky, Stebbins, Simpson, Mayr, Huxley, and Rensch in quilting population genetics together with other branches of evolutionary theory.

This grand union—epitomized by Julian Huxley's 1942 book *Evolution: The Modern Synthesis*—rested on the recognition and acceptance of two related conceptual pillars. The first was that gradual evolution could be explained by minute genetic changes which produce variation. This variation subsequently becomes the raw material upon which natural selection acts. The second tenet held that evolution at "higher" taxonomic levels and larger magnitudes could be explained by these same gradual evolutionary processes sustained over long periods of time. Thanks to the collective work of generations of biologists, the insights of Darwin and Mendel were finally reconciled, or so it seemed. The road ahead remained long and winding.

The forging of the Modern Synthesis of genetics and evolution is a story that has been recounted many times.[8] The relevant observation, from our perspective, is that, for all its ecumenical efforts, it remained incomplete:

> The Modern Synthesis established much of the foundation for how evolutionary biology has been discussed and taught for the past sixty years. However, despite the monikers of "Modern" and "Synthesis," it was incomplete. At the time of its formulation and until recently, we could say that forms do change, and that natural selection is a force, but we could say nothing about *how* forms change, about the visible drama of evolution as depicted, for example, in the fossil record. The Synthesis treated embryology as a "black box" that somehow transformed genetic information into three-dimensional, functional animals. (Carroll 2005, p. 7)

Embryology was treated as a "black box." To what is Carroll alluding?

Biologists had been aware of the intimate connection between development and evolution for quite some time. To wit, Darwin, T. H. Huxley, and their allies relied heavily on the—somewhat limited—embryological knowledge of their time as evidence for descent with modification, thereby phylogenetically connecting human beings to the rest of the animal kingdom. Given that both Darwin and Huxley correctly diagnosed development as the key to evolution, why was it left out of the Modern Synthesis?

The reason, Carroll maintains, is that for the century or so following the publication of Darwin's masterpiece, hardly any progress was achieved in understanding the principles of ontogeny. How a single unicellular zygote gradually unfolds to grow into a complex multicellular organism remained

---

[8] A classic recollection can be found in Mayr and Provine (1980).

a mystery, eluding any plausible biological explanation. As a consequence of this frustrating impasse, the study of embryology, heredity, and evolution—once inextricably tied at the core of biological thought—fractured, early in the twentieth century, into separate, independent fields. Each discipline strived to establish its own questions, principles, and methodology. This stalemate perdured for several decades. During this same time, embryologists became chiefly preoccupied with phenomena that could be studied by experimentally manipulating the eggs and embryos of a few species. As a result, the evolutionary framework, for them, faded into the background.

This black-boxing of the mechanisms of development was far from inconsequential for biological research at large. Evolutionists were studying genetic variation in populations, still largely ignorant of the relation between genes and form. Lack of embryological knowledge turned many serious scientific studies of the evolution of form into pointless speculative exercises. It also promoted the pernicious stereotype of evolutionary biology as a discipline better suited for dusty museums than research labs. How could we shed light on the evolution of form without having any firm empirical understanding of how said form is generated in the first place? Refining Mendel's original insights, population genetics had successfully established that evolution is grounded in changes in genes. Yet, as Carroll puts it, this by itself amounted to a "principle without an example" (2005, p. 8), as no gene affecting the form and evolution of any animal had been pinpointed, no matter how vaguely. In short, it gradually became clear that new insights in evolution would require breakthroughs in embryology.

Voices calling for a reunion of embryology and evolution, filling in the gaps left open by the Modern Synthesis, began to make themselves heard in the 1970s. A notable contribution came from Gould's 1977 book *Ontogeny and Phylogeny*, which revived discussions of how modifications of development may influence evolution. By this point, everyone was quite aware that genes are among the main common denominators of development and evolution. The problem was that little was known about *which* genes mattered in the development of organisms. This long-standing quandary was eventually solved by researchers working on *Drosophila*, the workhorse of genetics, who cleverly devised experiments to determine the genes that control the growth of fruit flies. And the discoveries were flabbergasting.

For over a century, biologists had assumed that various types of animals were constructed in completely distinct ways. If genes held the key to phenotypic variation, it was commonly supposed, then the greater the disparity in form, the fewer developmental processes two organisms would have in common at the genetic level. Things turned out very differently. Contrary to any expectation, most of the genes identified as governing the structure and organization of a fruit fly were found to have quasi-exact counterparts in most organisms, including humans. This was rapidly followed by another striking discovery. The development of various body parts—eyes, limbs, hearts, and the like—vastly different in structure across species and long assumed to have evolved in entirely distinct ways, was also governed by the same genes in different animals. So much for intuition!

All of this opened up an entirely new way of looking at evolution. The comparison of developmental genes across species spawned a new discipline at the interface of embryology and evolution: evolutionary developmental biology, or "evo-devo," for short. The discovery of this genetic toolkit, highly preserved across eras and phyla, provides further irrefutable evidence, in case additional proof was even needed, for the descent and modification of all organisms, humans included, from a common ancestor. Darwin's hypothesis was finally vindicated in full. This wealth of new data paints a vivid picture of how animal forms originate and evolve. And, perhaps even more importantly, it raises a host of new questions that will likely shape the future of much biological research for the years and decades to come.

The emergence of a common genetic toolkit and other analogies among species' genomes presents an apparent paradox, somewhat mirroring Darwin's own concerns regarding variability. If genes are so widely shared, how do differences arise? If species are so genetically similar, then why do they look so different? This puzzle was resolved by corroborating two related hypotheses. First, the development of form depends on the switching on and off of genes at different times and places in the course of development. Second, only a small fraction of DNA codes for protein—about 1.5%, according to recent estimates. Yet, ~3% is regulatory. Its function is to determine where, when, and how much of a gene's product is synthesized.

Time to start tying up some loose ends. Paraphrasing Carroll, we can conceive the story of evolution to date as a drama in three acts: Act One, Darwin's theory of evolution by natural selection; Act Two, the forging of the Modern Synthesis, which bridged Darwinism and Mendelism at various scales; Act Three, the ongoing Developmental Synthesis, spurred by the birth of evo-devo and other developments in molecular biology. All three acts star black boxes as protagonists, as a *leitmotiv*. At any point in time, crucial facts are unknown, misrepresented, or otherwise set to the side. There are questions that cannot be answered and hypotheses that cannot be verified. This ignorance does not stifle biological research; it enhances it. Borrowing a Kuhnian expression, these glitches are anomalies, not falsifications. As Mayr put it, Darwin wisely did not waste time and energy on problems then insoluble. He effectively postponed these issues until more auspicious times. Mendel followed suit and so did many other successful scientists. Unraveling the history of the life sciences requires keeping track of black boxes being set up and broken down. Our goal is figuring out how this works.

In conclusion, two points should be stressed. First, while the first part of this chapter focused on biology, black-boxing is by no means confined here. It permeates every area of science. Second, the tales of Darwin, Mendel, and their successors illustrate how black boxes play various crucial and subtly different parts in the advancement of science. The next two sections establish these claims by looking at two more cases.

§3.4. Absent-Minded: Psychological Behaviorism

Before the twentieth century, psychology was commonly viewed as the science of mental life, the systematic study of consciousness. Of course, Wilhelm Wundt, William James, Sigmund Freud, and many other prominent nineteenth-century psychologists realized that behavior is a form of evidence, and an important one for sure. However, none of these authors assumed that their entire discipline was directly aimed at predicting human conduct. To be sure, their approaches to the study of the mind presented significant methodological differences. For instance, James accepted physiological explanations as part of psychology and emphasized the descriptive power of viewing the mind teleologically. In contrast, the only form of causal explanation that Wundt allowed in his psychology was mentalistic causal explanation. Nevertheless, both agreed that conscious states could be explained by appealing to the laws governing mental life, and both presupposed that psychological differences would be relevant to such explanations.[9]

Early in the twentieth century, psychologists began to question the assumption that conscious states are their primary objects of study. It became increasingly popular to maintain that psychology concerns itself with

[9] The overview in this section draws on Flanagan (1991) and Hatfield (2002).

explaining and predicting human conduct and, more generally, the behavior of organisms.

Behavioral psychology comes in various forms and degrees of rigidity. Early proponents, such as William McDougall and Walter Pillsbury, advocated an account of behavior that unabashedly helped itself to the mentalistic vocabulary of traditional psychology. Thus, introspection was among the conceptual tools that could be employed in the process of explaining the mind. This proto-behavioristic science essentially introduced behavior as the chief object of study without eschewing mentalism *tout court*. More radical approaches rejected the validity of introspection for psychological analysis but retained the use of mentalistic terms in the description of human conduct. The most extreme proposals purported to expunge entirely all mentalistic talk from the vocabulary of psychology.

The most influential radical variant of early behaviorism was offered by John B. Watson. Watson was a hardcore materialist, strongly committed to the tenet that ultimate explanations of behavior would be grounded in the principles and language of physics and chemistry. Watson and his followers were well aware that psychology had ways to go before developing into a full-fledged behavioral science which eschewed all forms of intentional language. For this reason, they allowed the use of provisional explanations which charted stimulus-response patterns along the lines of Pavlov's conditioning theory and Thorndyke's laws of effects. Yet, this is where the line was drawn. Stimuli, responses, and other measurable bodily states were supposed to be rigorously described by using only the—allegedly—objective vocabulary of physical theories. In rejecting all inherently mentalistic notions, whether introspective or descriptive of behavior, psychology could finally join the ranks of legitimate natural sciences.

Psychology's path toward scientific respectability has become the subject of interesting debates in the history of science. According to mainstream reconstructions, this maturation was brought to completion when neo-behaviorists such as Tolman, Hull, and Skinner joined the methodological positivism of Schlick, Carnap, Hempel, and other thinkers associated with the "Vienna Circle."[10] On these traditional accounts, logical empiricism furnished psychology with the conceptual tools to rid itself of untestable metaphysical claims, thereby gaining rigor and objectivity. This was attempted by ensuring that descriptions of behavior and its triggers be translatable—that

---

[10] Classical accounts along these lines can be found in Boring (1950) and Leahey (1980).

is, reducible to—the language of physics along the lines of what eventually become the classical model, discussed in section 2.3 of Chapter 2.

Over the last few decades, this standard narrative of American psychology's alliance with positivism has been challenged. One revisionist account is founded upon the rejection of two assumptions presupposed, more or less explicitly, in the story rehearsed earlier.[11] From this standpoint, a first mistake is the belief that after psychology disenfranchised itself from philosophy, in the 1890s, a state of hostility existed between these two disciplines, until the rise of positivism reconciled them. A second misunderstanding lies in the failure to appreciate that virtually all neo-behaviorist psychologists rejected the main conceptual pillars of logical positivism, such as verificationism and the analytic-synthetic distinction. This does not mean that philosophy had no role to play in the development of behaviorism. Quite the contrary, philosophers were involved from the very beginning, and such discussions played a formative role in the maturation of Hull's and, especially, Tolman's thought. This revisionist reconstruction rejects the common view that philosophy of science only dawned in North America when positivists migrated, following the surge of Nazism across Europe. Mainstream American philosophical traditions—such as realism, neo-realism, and critical realism—had long called for philosophy to take the sciences seriously. In short, from this perspective, the actual relations between philosophy and psychology in American academia were already reasonably amiable, promoting a fruitful spirit of intellectual exchange.

Adjudicating between historical narratives lies beyond my interest and professional competence. Be that as it may, let us fast-forward to the end of the story and take a look at the outcome of this revolution in psychology.

The most developed form of mature behaviorism comes from the work of B. F. Skinner, arguably the most influential psychologist to ever live and work in the United States. Constructing psychology as the science of behavior, in stark opposition to previous characterizations as the study of mind or consciousness, Skinner followed the trail blazed by early behaviorists. Especially influential on the young Skinner was Watson, who had diagnosed the delayed growth of psychology as the effect of pervasive metaphysical and epistemological vices. These attitudes were readily identified as a nefarious legacy of Descartes's substance dualism, which implied that minds are nonphysical, private, unobservable entities, essentially rendering any serious

---

[11] The following reconstruction is due to Amundson (1983, 1986) and Smith (1986).

scientific approach to psychology a chimera. Yet, as we shall see, Skinner's mature views were, in many important respects, more nuanced and sophisticated than the ones of his illustrious behaviorist predecessors.

The attempt to place psychology on more secure methodological grounds was undoubtedly a noble endeavor. However, in decisively rejecting any form of mentalistic vocabulary and confining the study of the mind to talk of overt conduct, early forms of radical behaviorism ended up trivializing psychology by providing a drastically impoverished reconstruction.

In the initial stages of his long and productive career, a phase pervaded by the intellectual milieu of operationalism, Skinner seems to have fallen prey to precisely this same mistake. Yet, by the end of World War II, he had clearly realized that an uncompromising psychological theory, which eschews as meaningless all subjective, cognitive, and affective phenomena, would fall short of a plausible, adequate, and comprehensive theory of mind.[12] Like his behaviorist precursors, Skinner set out to replace Cartesian dualism with a programmatic materialist stance according to which all mental events are physical and, therefore, they must obey the same laws that govern overt behavior. But, contrary to Watson's relatively simplistic reductive materialism, Skinner's behavioristic psychology is much more nuanced and complex. The fundamental difference between mental and physical phenomena, for Skinner, is not a straightforward ontological matter. Behaviorism is an epistemic stance. Allow me to elaborate.

To see what exactly is at stake, consider one of Skinner's mature publications: *Verbal Behavior* (1957). This volume, which systematizes the results of over two decades of extensive experimental work, aims to provide a systematic and comprehensive functional analysis of linguistic behavior. What Skinner is after is a precise specification of the interactions between the variables that control human action and, specifically, those which determine particular verbal responses. These variables ought to be described exclusively in terms of physical stimuli, reinforcement, deprivation, and other notions that Skinner and his collaborators have painstakingly operationalized over numerous experimental studies of animal conduct. As Chomsky (1959, p. 26) states in his scathing review of this monograph, "the goal of this book is to provide a way to predict and

---

[12] Skinner is often portrayed as advocating a naive "stimulus-response psychology," analyzing behavior in terms of unconditioned and conditioned reflexes, along the lines of Pavlov and Watson. This, however, is a mischaracterization, as witnessed by Skinner's attempt to integrate traditional S-R models with the concept of operant behavior.

control verbal behavior by observing and manipulating the physical environment of the speaker."

What makes Skinner's project so ambitious and controversial? Contrary to common misconceptions, the answer is neither that it sets out to provide a functional analysis of behavior, nor that it restricts itself to studying observables per se. The proposal is so radical, Chomsky notes, because of the specific limitations on the observables under investigation and, in particular, due to the simplistic description of the causes of behavior.

> Skinner's thesis is that external factors consisting of present stimulation and the history of reinforcement (in particular the frequency, arrangement, and withholding of reinforcing stimuli) are of overwhelming importance, and that the general principles revealed in laboratory studies of these phenomena provide the basis for understanding the complexities of verbal behavior. He confidently and repeatedly voices his claim to have demonstrated that the contribution of the speaker is quite trivial and elementary, and that precise prediction of verbal behavior involves only specification of the few external factors that he has isolated experimentally with lower organisms. (1959, pp. 27–28).

One can perhaps paraphrase Chomsky's insight along the following lines. What makes Skinner's project controversial is his deliberate choice to omit any reference to structures "internal" to the organism, the structures which govern the relation between inputs and outputs. Whether stimuli and environment play a role in shaping behavior is not in question—of course they do. And, in the absence of neuropsychological data, they might well be the main or even the only kind of evidence.[13] The striking move is the complete rejection of abilities and capacities germane to the organism itself.

These considerations highlight how Skinner's strategy mirrors the approaches of Darwin and Mendel. All these authors effectively set aside the details of the mechanisms producing the behavior in question. Many readers will find this perplexing. Behaviorism might survive, in various revised

---

[13] Interestingly, Chomsky is adamant on this score: "Insofar as independent neurophysiological evidence is not available, it is obvious that inferences concerning the structure of the organism are based on observation of behavior and outside events. Nevertheless, one's estimate of the relative importance of external factors and internal structure in the determination of behavior will have an important effect on the direction of research on linguistic (or any other) behavior, and on the kinds of analogies from animal behavior studies that will be considered relevant or suggestive" (Chomsky 1959, p. 27).

forms, as a methodological directive in psychology, for instance, in applied behavioral analysis therapies. But is it not an obsolete theory of mind? How can it be compared to some of the most successful theories in all of science? Let me reassure everyone. I am not suggesting that these theories are on a par. The parallel is simply meant to show that Skinner's controversial move cannot be his decision to construct a black box. What is problematic is the nature and structure of the black box itself, that is, the specifics of what is included in and what is omitted from the explanation.[14]

Time to tie up some loose ends. The behaviorist attempt to make psychology respectable on scientific grounds involved a systematic attempt to expunge mental states and other internal features from explanations of human conduct. This, it is often claimed, was Skinner's capital sin: black-boxing mental states. This conclusion cannot be correct. Now, surely, "analytic" or "logical" behaviorism is dead, at least as a philosophical analysis of the nature of mental states. Let it rest in peace. At the same time, as we saw, Darwinian evolutionary theory and Mendelian genetics also contain black boxes, and so do molecular biology and evo-devo. Since these theories are far from discredited, black-boxing per se cannot be the root of the trouble for behaviorism. What did Skinner do that Darwin and Mendel managed to avoid? Where exactly did behaviorism take a wrong turn?

The key to answering these questions is to note that not all black boxes are created equal. Not all processes and events are black-boxed for the same reasons, and some attempts are more successful than others. Darwin and Mendel were unable to accurately depict the mechanisms of inheritance and variation. The completion of the Modern Synthesis had to await the breakthroughs of evo-devo. Similarly, Skinner and his followers lacked substantive knowledge of the neuropsychology of behavior. Still, as Chomsky lucidly points out, this is only part of the story and not the controversial one. The behaviorist rejection of mental states is not merely an admission of ignorance like we saw in the biological cases. The black-boxing of internal psychological variables is a deliberate methodological strategy to eschew their relevance in the causal explanation of action. Talk of mental states should be avoided, within a Skinnerian perspective, not because such entities do not exist or because not enough is known about them. By reconstructing psychology in such a way as to avoid any reference to mental events and neural

[14] As Skinner (1974, p. 233) remarked in a later book: "The organism is, of course, not empty, and it cannot be adequately treated simply as a black box, but we must carefully distinguish between what is known about what is inside and what is simply inferred."

processes, Skinner purported to dismiss—or, more charitably, to minimize—their relevance in causal accounts of human conduct. In this sense, the mind is completely absent from his psychological theory.

There are two general morals that can be drawn from this excursus into the history of psychology. First, behaviorism provides another clear example of the ubiquity of black-boxing across the sciences. Mental states are explicitly treated as black boxes: they are complex phenomena that, for the sake of explaining human conduct, can be replaced by patterns of inputs and outputs, by stimuli-and-reaction models. In this respect, Skinner's treatment of the mechanisms of behavior resembles Darwin and Mendel's functional description of the mechanisms of inheritance and variation.

Second, and more important, the methodological insulation of mental states reveals an aspect of black-boxing that was sidelined in the biological examples. Darwin and Mendel avoided any discussion of the mechanisms of variation, heritability, and ontogeny because of ignorance and incorrect knowledge. The behaviorist strategy, in contrast, eschews any direct, unmediated appeal to mental states because they were deemed unsuited to be investigated scientifically. The difference could hardly be more prominent.

These observations raise a host of follow-ups. How was Darwin able to set aside the heart of his theory without compromising the structure and success of his explanations? How could Mendel found the field of genetics while having virtually nothing to say about the processes by which traits are passed on from parent to offspring? What was Skinner's mistake in setting mental states to the side? Could it have been avoided? These important issues require a more systematic discussion of black-boxing. I will undertake this project in the ensuing chapters. Before doing so, let me provide one final example, which will unveil additional layers of complexity. After all, my immediate aim is to convince you that black boxes can be much more diverse and systematic than contemporary scholars typically suppose.

§3.5. The Nature and Significance of Economics

What is economics? In Chapter One of his *Essay On the Nature and Significance of Economic Science*, Lionel Robbins proposes one of the most influential modern definitions: "Economics is the science which studies human behaviour as a relationship between ends and scarce means which have alternative uses" (1932, p. 16). The fundamental axioms of the enterprise of economics, Robbins maintains, are grounded in a theory of value which, in turn, is based on a theory of preference. In this account, then, the whole of economics rests on the capacity of agents to arrange their preferences in some order.

This construction of economics on mental foundations might seem puzzling to contemporary readers. Prima facie, Robbins's definition turns economics into a subfield of psychology, namely, the study of how human agents rank available options, given constraints on limited temporal and material resources. Robbins is quick to address the concern. *Why* and *how* creatures like us attach particular values to specific objects of choice are substantial empirical problems. Yet, he glosses, these issues do not concern economists. They are questions for psychology. All that economists—*qua* economists[15]— need to assume is the ranking of options itself.

> Why the human animal attaches particular values in this sense to particular things, is a question which we do not discuss. That is quite properly a question for psychologists or perhaps even physiologists. All that we need to assume as economists is the obvious fact that different possibilities offer different incentives, and that these incentives can be arranged in order of their intensity. (Robbins 1932, p. 86)

This is an adage that most students of economics learn at an early stage. Still, in the ensuing discussion, Robbins adds a remark that many of his colleagues would now find disconcerting. No economic explanation, he claims, can be considered adequate, let alone complete, without invoking elements of a subjective or psychological nature, such as preference.[16]

This ineliminable subjective component in economic explanations raises issues. From a historical perspective, it strikingly contrasts with the methodological orthodoxy of the day—logical empiricism—which purported to

---

[15] This disclaimer is required because nothing prevents the same individual or research team from pursuing agendas in both economics and psychology. Still, for Robbins's point to go through, it is enough to distinguish psychological *questions* from economic ones.

[16] "But even if we restrict the object of Economics to the explanation of such observable things as prices, we shall find that in fact it is impossible to explain them unless we invoke elements of a subjective or psychological nature. It is surely clear [. . .] that the most elementary process of price determination must depend *inter alia* upon what people think is going to happen to prices in the future. [. . .] It is quite easy to exhibit such anticipations as part of a general system of scales of preference. [Footnote omitted.] But if we suppose that such a system takes account of observable data only we deceive ourselves. How can we observe what a man thinks is going to happen? It follows, then, that if we are to do our jobs as economics, if we are to provide a sufficient explanation of matters which every definition of our subject-matter necessarily covers, we must include psychological elements" (Robbins 1932, pp. 88–89).

reduce all synthetic knowledge to statements that are testable, at least in principle. Mental states are not directly observable. Thus, they cannot be tested, in practice or in theory. According to the verificationist semantics endorsed by logical positivism, this turns them into meaningless gibberish.

The methodology of economics vis-à-vis its status as a science quickly became a contentious matter requiring serious discussion. Prominent scholars, such as Terence Hutchison (1938), maintained that economics, as a bona fide science, must formulate falsifiable generalizations and subject them to rigorous test. This requirement might initially appear plausible, even truistic. Upon further scrutiny, it poses strictures of all sorts. Conforming economic methodology to the basic tenets of logical positivism turned out to be a real challenge, independently of its appeal to psychological elements.

For one thing, several mainstream economic assumptions are heavily hedged with *ceteris paribus* qualifications and, hence, not directly testable. The reason is straightforward. An unqualified generalization such as "cutting taxes is always followed by a raise in employment rates" can be corroborated by ensuring that the consequent, a raise in employment rates, invariably follows the antecedent, a cut in taxation rates. In contrast, the hypothesis that "all things being equal, cutting taxes is followed by a raise in employment rates" is harder to assess. When exactly should one deem it falsified? When should things be considered equal and by what standards?

A second and more troubling issue emerged when empirical economic claims were actually tested. Many turned out to be false. For instance, according to a well-known theoretical hypothesis, economic agents operate under the assumption of marginal cost pricing. However, a survey was published revealing that real-world firms actually base their calculations on full-cost pricing. A few years later, another study showed that firms do not maximize profits by equating marginal costs and marginal revenues, another core tenet of early neoclassical economics.[17] In short, it soon became evident that economic models provide an inaccurate depiction of business conduct. If economists were supposed to behave like responsible scientists—which, in the logico-positivist intellectual milieu of the early twentieth century, meant responsible physicists—they were not doing a very good job.

This impasse triggered various kinds of responses. Economists associated with the Austrian School—such as Frank Knight, Carl Menger, Ludwig von Mises, and Friedrich Hayek—willingly accepted the conclusion that the

---

[17] These results are reported by Hall and Hitch (1939) and Lester (1946), respectively.

standards of natural science simply do not apply to economics. This, how-ever, is a hard bullet to bite. What methodology should economics adhere to, if natural science is not the answer?

A different route involved showing that, despite superficial differences, economics did meet the stringent positivist criteria. How? One option for salvaging economics in the face of the falsity of core assumptions was to treat its basic laws as truths, albeit *inexact* truths. Unfortunately, from the standpoint of logical empiricism, this approach amounts to methodolog-ical suicide. If the benchmark of rigorous scientific statements lies in their testability, inexact truths are just as hard to verify as hedged *ceteris paribus* generalizations. How is an inexact truth confirmed? When is it discarded, and what should replace it? All these solutions appeared to be non-starters. A more promising strategy was soon to be offered.

The way out of the dilemma was provided by Milton Friedman, in his cel-ebrated 1953 article, "The Methodology of Positivist Economics." Friedman argued that the truth or falsity of economic assumptions is perfectly irrel-evant to the acceptance or rejection of a positive scientific theory. The only thing that matters is whether the hypothesis is predictive and explanatory with respect to the class of phenomena that it is intended to cover. An ade-quate theory, Friedman goes on to clarify, is one that explains much by little. It must abstract the crucial factors from the mass of complex, idiosyncratic circumstances surrounding the phenomena to be explained, and predict fu-ture states on the basis of these core elements alone. Hence, the relevant ques-tion to ask about the assumptions of a scientific theory is not whether they are descriptively realistic. They never are. The appropriate issue is whether our hypotheses constitute good enough approximations for the purposes at hand, whether they yield accurate predictions.

Friedman illustrates his contention with simple and effective examples. Consider the law of free-falling bodies in elementary physics: $s = 1/2gt^2$, where s is the distance in feet and t is time in seconds. Take the hypothesis that leaves are positioned around a tree as if each leaf deliberately seeks to maximize its exposure to sunlight, relative to the position of its neighbors. Or the claim that expert billiard players make their shots as if they computed the complicated mathematical formulas that specify the optimal direction of travel. Taken literally, all these hypotheses are obviously false. The preceding formula only captures the falling of bodies in a perfect vacuum. Leaves do not seek to maximize anything, let alone sunlight exposure. And no human being comes remotely close to computing in real time the equations underlying the motion of billiard balls. Nevertheless, Friedman argues, these hypotheses are commonly accepted. This is because they allow us to predict, with reasonable accuracy, their target phenomenon: the falling of dense objects, the distribution of leaves on trees, and the outcome of billiard shots. More precisely, Friedman argues, these hypotheses are typically accepted because they strike the right balance between precision and complexity. Sure, more predictive generalizations are available. Yet, in most contexts, this greater predictive accuracy is offset by the required amount of calculation. Could one not say the same thing about economic tenets, such as the presupposition that individuals and firms behave as if they sought to maximize their effective returns? Economists accept or reject assumptions because of their predictive power, not because of their (in)accuracy.

Few methodological directives have been as widely debated as the one just presented. And yet, Friedman's argument is often mischaracterized. First, some have countered that realism is not detrimental to hypotheses.[18] But, on a charitable reading, this is not Friedman's contention. The point is rather that realism should not be a factor in theory choice. One can always prefer a more accurate hypothesis to a less realistic one, without loss of explanatory power, as long as the former has a better predictiveness-to-complexity balance than the latter. Regardless, the justification for doing so resides in the greater predictive value of the hypothesis, not in its realism.

Second, Friedman's strategy conforms economic methodology to the standards of logical empiricism.[19] Positivistic tenets such as the in-principle testability of all scientific statements have long been rejected. Does this not, *ipso facto*, make Friedman's argument obsolete? The answer is negative. Friedman's contention stands or falls independently of radical forms of empiricism. In addition, this pioneering directive anticipates contemporary discussions of the role of abstractions, idealizations, and other forms of deliberate misrepresentation, which have now finally found their way into mainstream philosophy of science.[20] In short, Friedman's strategy

---

[18] See, for instance, Simon (1963).

[19] The connection between Friedman and positivism are clearly seen in Nagel (1963), who maintains that, while Friedman's argument lacks cogency, it is nevertheless sound.

[20] The use of fictions and distortions is commonly assumed to be a long-standing scientific practice. This, as argued by Chomsky (2012), is a misconception. The idea of taking a complex system and transforming it into an artificial construct, which can be investigated in depth, was a huge step that was only taken in relatively recent times. Galileo saw laws principally as calculating devices. It wasn't until the twentieth century that theoretical physics became a legitimate field in itself. To this day, Chomsky notes, the idea of constructing idealized systems for the sake of computational tractability is considered somewhat exotic outside of the hard-core natural sciences, in fields such as linguistics.

is more powerful than most critics realize. No wonder it has been so influential.[21]

The final ingredient required to reconcile Robbins's dictum—the ineliminable presence of subjective psychological components in economic explanations—with logical positivism is the *revealed preference view of utility*. By assuming that choice and preference exhaustively determine each other, effectively making the two notions interchangeable, preference may continue to perform an essential role in economics without positing intangible mental quantities. Identifying preference with choice, the former essentially becomes as objectively measurable and testable as the latter. Economics could abide to positivist demands, after all.

Since the 1950s, mainstream neoclassical economics has come under fire from various standpoints. First, famous paradoxes due to the independent work of Allais and Ellsberg have unveiled widespread and systematic violations of the axioms of expected utility theory.[22] Second, the "definitional egoism" underlying rational decision-making has been critiqued as an oversimplified view of human choice. Third, neoclassical economics is said to lack normative force and also to give up on the intuitive assumption that mental states explain behavior.[23] I shall set all these critiques aside. My goal is not to

For an insightful philosophical discussion of idealization—which will also be discussed more systematically in Chapter 5 and Chapter 7—see Appiah (2017).

[21] I should clarify that my sympathetic assessment of Friedman's strategy deliberately focuses on his general methodology and abstracts away from his substantive ideological baggage, political project, and underlying agenda. As George DeMartino (2000) has convincingly argued, the crusade led by Friedman and his disciples in defense of the free market depended heavily on the assumption that the preferences emerging within a free market are legitimate. This, in turn, presupposes that preferences are exogenous to economic activity. The rationale is simple. Within neoclassical economics, the market is the primary means of preference satisfaction. But if the market also shapes, influences, or otherwise warps the preference that it satisfies, then we cannot point to preference satisfaction as a defense of free markets. Neoclassical theory sidesteps the problem by presuming exogenous preferences. Similarly, neoclassical economists did not invariably, or even reliably, practice what they preached. While holding up the test of prediction to validate their own unrealistic assumptions, the profession also became intolerant of all other approaches. Not only were alternatives not promoted. In addition, neoclassicists occasionally actively thwarted attempts to show that different methodologies did better by their own standards of predictive performance. To wit, much econometrics in the twentieth century was directed toward vindicating the claim that all would benefit from free trade, that minimum wage actually harmed people more than it helped, and other items on the agenda of neoliberalism. Doing justice to these considerations requires a longer—and valuable—discussion that, however, transcends my aim and competence. Still, I stress that my praise of Friedman's methodological strategy does not imply any endorsement of sociopolitical agendas or exploitation of double standards when it comes to choice and implementation of theory.

[22] As we shall see, in greater detail, in section 6.5, expected utility—the utility of an entity or aggregate economy over a future period of time, given unknowable circumstances—is standardly employed to evaluate decision-making under uncertainty.

[23] A classic critique of "definition egoism" is Sen (1977). Philosophical discussions concerning normativity (or lack thereof) are found in Hausman (1992) and Rosenberg (1992).

assess neoclassical economics *tout court*. Rather, I wish to bring to light some methodological presuppositions of contemporary economics.

One observation is especially important from our perspective. As noted, the combination of Friedman's "as if" approach with the revealed preference view of utility effectively made economic theory kosher from a positivist standpoint. Yet, in doing so, it also completely shielded economic models from psychological influence and critique. The rationale is simple. In neo-classical economics, preference is not something that is open to further analysis. The identification of preference with choice turns both into raw data that explain the behavior of a rational—that is, a logically consistent—agent. In a nutshell, choice, preference, and utility are all black boxes.

This screening off of psychology had a triple effect on twentieth-century economics. First, identifying preference with choice provided a clear, simple, and operational way to talk about reasons and preferences. Complications arising from the idiosyncrasy of human psychology and the neural implementation of these processes, both of which, at that time, were far beyond our ken, could be unabashedly swept under the rug. Second, if, from an economic perspective, all there is to preference is revealed choice, then psychoneural discoveries about the human motivational system, in principle, cannot shed any light on economics. We can effectively talk about behavior without mentioning its causes. In addition, and more radically, specifying such causes will leave our understanding of economic behavior untouched. Third, the isolation of economics from psychology was a powerful and effective strategy to screen off the core assumptions of neoclassical theory from potential attacks. Specifically, the superiority of free markets could not be questioned by appealing to the psychological states of economic agents. One could not argue, for instance, that the preferences that led to choice were somehow normatively deficient.

The insulation of economics might have been relatively inconsequential during the days of logical positivism, when psychology and neuroscience were still relatively young, underdeveloped, and had little to contribute to the study of economic phenomena. To be sure, the critical examination of the relation between preference and behavior has a long history, harking back to Plato and Aristotle. Still, until recently, we lacked any systematic theory which might shed light on how these choice-mechanisms are actually implemented in the human mind, allowing for more accurate predictions and sharper explanations. Things have changed quite drastically over the last few decades. Following the cognitive revolution, proposals regarding the impact of psychology and neuroscience on economics have emerged.

The claim underlying contemporary "psycho-neuro-economics" (PNE) is that new insights into how minds and brains actually frame, compute, and resolve problems challenges fundamental assumptions regarding the behavior of economic agents.[24] For instance, behavioral economics has denounced the lack of reference points in expected utility theory. In doing so, it has emphasized how real humans, as opposed to economic agents ("econs"), use heuristics and biases in making decisions, rather than calculating expected utilities. Neuroeconomics has also made some progress revealing how and where mechanisms governing choice are implemented in the brain, unveiling further inconsistencies with conventional wisdom. In short, the story goes, classical economic notions, such as risk aversion, time preference, and altruism, are due for a makeover. Is it finally time to open the black boxes that were so skillfully crafted by Robbins, Friedman, and their colleagues?

Supporters of PNE answer affirmatively. This is clearly stated in one of the founding manifestos of neuroeconomics (Camerer et al. 2005, p. 53):

> Economics parted company from psychology in the early twentieth century. Economists became skeptical that basic psychological forces could be measured without inferring them from behavior (the same position "behaviorist" psychologists of the 1920s reached), which led to adoption of the useful tautology between unobserved utilities and observed (revealed) preferences. But remarkable advances in neuroscience now make direct measurement of thoughts and feelings possible for the first time, opening the "black box" which is the building block of any economic interaction and system—the human mind.

It is worth stressing that, in calling classical economic notions "black boxes," the claim is not that these notions were unstructured. Von Neumann, Morgenstern, Friedman, and their colleagues had much to say about the nature of choice, preference, and utility. The point is, rather, that these neoclassical economic constructs are black boxes relative to the underlying

---

[24] For the sake of convenience, I introduce the moniker "psycho-neuro-economics" to lump together various contemporary developments in behavioral economics and neuroeconomics. Whenever the need for a finer-grained classification arises, I draw subtler distinctions between behavioral economics proper and its neuroeconomic counterpart.

neuro-psychological mechanisms, which played no role, in principle or in practice, in determining and explaining their constitution.

Not everyone, however, is so enthusiastic. Naysayers vary with respect to the scope and depth of critique. Some acknowledge the potential of neuroeconomics and, yet, complain that, so far, it has delivered a mixture of marketing bloopers, bad epistemology, and messy experimental practice.[25] Others fear that outsourcing the study of preference to psychology may alter the very nature of economic explanations, turning them into questions about how the mind works. In what follows, I focus on a different worry.

A more trenchant objection, which echoes Friedman from a current standpoint, maintains that psychology and neuroscience, in principle, are irrelevant to economics.[26] Economics and psychology, from this perspective, address distinct questions, have different objectives, and introduce their own abstractions. Moreover, contrary to the therapeutic aims of PNE, neoclassical theory identifies welfare with choice without purporting to advance a theory of happiness. On this view, opening the black box of mental states does nothing to economics because it is not part of economics at all.

In conclusion, the history of economics illustrates another role for black-boxing in science. By isolating a class of phenomena and insulating it from further explanation, one can draw disciplinary boundaries between subfields. These boundaries can then be used to advance a research project in the face of epistemic limitations—it allows us to define and employ a concept when not much is known about its underpinnings. Alternatively, black-boxing can also become a strategic way to shore up and protect the most vulnerable components of a theory against its rivals. A black box, in this sense, acts as a protective belt. Perhaps classical and early neoclassical economic theory segregated mental states because of ignorance or to render their disciplines fully testable and, thereby, scientific. But no contemporary economist is unaware of the staggering discoveries regarding psycho-neural mechanisms accomplished over the last few decades. No scholar worth their salt would dismiss core studies of the mind and brain as pseudo-scientific. It is the relevance of these discoveries for economics

---

[25] Harrison (2008) humorously dubs this programmatic approach "MacGyver Econometrics." Indeed, even authors who are more sympathetic to PNE than Harrison, such as McCabe (2008), will admit that the contribution so far has been somewhat limited.

[26] The main reference here is an influential article by Gul and Pesendorfer (2008).

that is at stake. PNE maintains that understanding these mechanisms is advancing economic theories. This point of view is expressed clearly by Camerer et al. (2005, p. 10): "The 'as if' approach made good sense as long as the brain remained substantially a black box. The development of economics could not be held hostage to progress in other human sciences." In contrast, advocates of a "mindless" form of neoclassical economics, which views economics and psychology as fundamentally distinct, question the relevance of these findings for their own discipline: "Our conclusion is that the neuroeconomic critique fails to refute any particular (standard) economic model and offers no challenge to standard economics methodology" (Gul and Pesendorfer 2008, p. 7).

Who wins this battle? While I shall not take sides on this issue, I want to stress that the current disagreement hinges, to a large extent, on the characterization and methodological boundaries of economics. If economics, by definition, presupposes the choices of agents, then psychology and neuroscience clearly have no relevance. The opposite conclusion can be drawn on a broader conception, that defines economics as the study of a special class of choices.[27] We shall return to this debate in Chapter 6. The point, for the time being, is that this duality is precisely what we should expect given the situation discussed in section 2.6 of Chapter 2. PNE embodies the reductionist tenet that breaking down economic explanations to more fundamental constituents increases their power. Neoclassical economics, in contrast, embodies the antireductionist tenet of autonomy. However, as currently presented, the dispute is generally terminological. We need a different approach to make the discussion more substantial. The key, I will argue in later chapters, is understanding utility and other economic constructs as black boxes.

---

[27] To wit, an influential neuroeconomist, Camerer (2010, p. 43), offers a broad definition of economics as "the study of the variables and institutions that influence economic choices, choices with important consequences for health, wealth, and material possession and other sources of happiness." While Gul and Pesendorfer (2008) do not offer an explicit comprehensive definition of economics, they narrow the field substantially with assertions such as "standard economics focuses on revealed preference because economic data comes in this form" (p. 8). It should be evident that, on the former definition, it becomes hard to dismiss the relevance of psychological discoveries on economic choices. In contrast, the latter delimitation virtually dismisses the import of economics on psychology by definition, by severely constraining what may and may not count as economic data.

§3.6. The Many Faces of Black Boxes

The practice of successfully explaining a phenomenon despite explicitly appealing to entities and processes the details of which are left unspecified or mistakenly identified has a long and hallowed history. Newton's treatment of gravity—culminating in his famous dictum "Hypotheses non fingo"—is a fascinating example. Yet, I focused our present discussion on four examples from three special sciences: biology, psychology, and economics.

Our excursus into the history of science began by tracing its steps back to the roots of contemporary evolutionary thinking. Darwin's induction ceremony into the Scientists' Hall of Fame is sacrosanct. He is the principal founding father of one of the great success stories in modern biology: the theory of evolution by natural selection. At the same time, it comes as no surprise that much of what we refer to as "evolutionary theory" today, including what is standardly called "Darwinism," is absent from Darwin's original writings. For instance, the great English naturalist is candid in the *Origin* about his ignorance regarding the mechanisms of inheritance and variation. And his subsequent speculative hypotheses purporting to fill in these blanks were strikingly off the mark. Despite these evident shortcomings, Darwin's evolutionary explanations are far from deficient or misguided. How is this possible? The key lies in Darwin's deliberate decision to black-box the processes underlying inheritance and variation. Their central role in evolutionary theory was crystal clear. Yet, a systematic exploration of their nature and structure had to be set to the side until much more was figured out.

My second illustration of black-boxing "in action" focused on Mendel's groundbreaking findings, which effectively treated genes as black boxes. These results, rediscovered in 1900, were finally incorporated into mainstream biology and brought to bear on phylogeny in the first half of the twentieth century, with the forging of the Modern Synthesis. The grand union of genetics and evolution effectively opened the black boxes of heredity and variation, answering many questions left unanswered or inadequately addressed by Darwin and Mendel. At the same time, it constructed new black boxes, for instance, segregating the mechanisms of ontogeny. Issues set aside by the architects of the Modern Synthesis have recently begun to be unraveled, in the context of the so-called Developmental Synthesis. Along the same lines, early molecular biology successfully opened Mendel's black box, the biochemical structure of genes, only to replace it with new placeholders. With the steady growth of molecular techniques, the details underlying the regulation of gene expression and the principles governing the new "omics" fields in biology, initially left untouched, started to emerge.

The third case history explored the history of psychology. Until the nineteenth century, psychologists typically considered mental states to be private conscious phenomena not amenable to public scrutiny. In an attempt to endow their discipline with a stricter scientific methodology, behaviorists proposed that psychology restrict itself to seeking laws linking stimuli to behavior. The rationale was that only what is publicly observable is a fit subject for science, essentially excluding mental states, as traditionally conceived, from rigorous scientific examination. From our perspective, the black-boxing of mental states enacted by behaviorism is interesting for two reasons. First, it provides yet another clear and historically significant instance of the widespread use of black boxes in scientific practice. Second, and more important, this methodological segregation of mental states highlights some aspects of the general strategy that were not evident in the biological cases.

My final example came from the field of economics. In eighteenth- and nineteenth-century Britain, economics and psychology were considered two branches of a single subject: moral philosophy. Understanding the behavior of markets presupposed a reasonably accurate psychological portrait of individual agents. It is thus unsurprising that seminal works by authors such as Smith, Mill, Edgeworth, Marshall, and Pigou included discussions of how mental states affect economic interactions. With the development of neoclassical economics, however, feelings and other psychological states came to be set aside, effectively isolating economics from fields such as psychology and, later, neuroscience. Why did this occur? Simply put, feelings came to be perceived as useless constructs, meant to predict behavior, but which could only be inferred from behavior itself. In the 1940s, the concepts of ordinal utility and revealed preference eliminated the superfluous intermediate step of positing unmeasurable feelings, equating unobserved preferences with observed choices. The risk of circularity was avoided by requiring consistency in behavior. Once an agent reveals a preference for *a* over *b*, they ought not subsequently choose *b* over *a*. This made the theory falsifiable, conforming economics to positivist strictures. This "as if" approach makes perfect sense as long as the brain is considered a black box. Yet, following almost a century of separation, economics has begun reimporting insights from psychology. Behavioral economics is now a prominent fixture on the landscape and has spawned applications to various areas of economics, such as game theory, labor economics, public finance, law, and macroeconomics. Assessing current debates regarding whether and how psychological and neuroscientific discoveries can benefit economic theorizing requires understanding the black-boxing of mental states during the neoclassical revolution. Are these black boxes really in the process of being reopened? Should they be?

The case studies presented in this chapter play a dual role. First, they illustrate the prominence of black-boxing across fields as diverse as evolutionary theory, genetics, evo-devo, behavioral psychology, and economics. One could say that the history of science is a history of black boxes, continually unwrapping some, while packing others (Figure 3.1). Second, they show that, contrary to a common if tacit presupposition, black-boxing is not a one-trick pony, a monolithic strategy that merely allows scientific research to proceed in the face of our ignorance. Our excursus highlights that black boxes can play various, subtly different roles across disciplines.

Darwin and Mendel set aside the principles of inheritance and variation because they lacked accurate knowledge of the key mechanisms. This strategy effectively postponed important issues that were intractable in their day and age. Similarly, evo-devo's ongoing attempt at cracking development requires theoretical and experimental breakthroughs that had not yet emerged during the forging of the Modern Synthesis. In all these cases, black-boxing is a way of masking our ignorance for the sake of progress.

The maturation of psychology between the late 1800s and the late 1950s tells a different tale. Behaviorism is an attempt to reconstruct an old and venerable discipline on a more secure scientific foundation. This is a noble endeavor. The primary challenge that confronted Watson, Skinner, and their

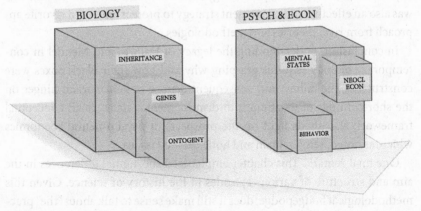

Figure 3.1. The history of science is a history of black boxes.

followers was not overcoming lack of psychological knowledge. Their methodological crusade was directed at a different target. They chiefly wanted to show how it is possible to engage in an empirical study of the mind without appealing to private, inaccessible, and unmeasurable properties. This require bracketing mental states, as traditionally conceived. Contrary to Darwin and Mendel, however, these black boxes were not meant to be set aside until later times, when they could finally be opened. Behaviorism used the black-boxing strategy as a trash can to dispose of concepts that would ultimately muddle a rigorous, scientific study of the mind.

From a philosophical perspective, behaviorism can be viewed as an application of Popper's demarcation problem to a developing field, with the aim of distilling a truly scientific methodology purged of pseudo-scientific lures. The history of economics emphasizes how black-boxing can also serve a different—but equally important—purpose: drawing boundaries between disciplines, thereby protecting and insulating a field from "external" critique. Despite its flirtatious relation with behavioristic psychology and positivist philosophy of science, early neoclassical economics had little interest in dismissing the study of mental properties as pseudo-scientific. The central contention was that psychological states are not the appropriate subject matter for economics. Allow me to illustrate the difference with a somewhat colorful metaphor. Behaviorism treated the black box containing mental states as a sort of Pandora's Box that is better off hidden to avoid dangerous temptations. Neoclassical economists, in contrast, considered that theoretical construction as a housewarming gift to be repackaged and shipped across campus to their colleagues working in revamped or newly founded departments of cognitive psychology and neuroscience. Incidentally, this was also an effective and convenient strategy to protect one's own favorite approach from rival theories and methodologies.

In conclusion, understanding the legacy of Darwin and Mendel in contemporary biology requires grasping why and how their black boxes were constructed. The failure and consequent demise of behaviorism hinges on the shortcomings of its attempt to dismiss mental states. Such conceptual framework also sheds light on the prospects of psycho-neural economics which, as noted, remain open and hotly debated issues.

One final remark. This chapter emphasized substantial differences in the aim and structure of various episodes in the history of science. Given this methodological hodgepodge, does it still make sense to talk about "the" practice of black-boxing in the singular? This entire book is based on the premise that yes, it does make sense. Setting important divergences aside, all these re-markable case studies from the history of science converge on the same core phenomenon. This is the identification of some principle that, for various reasons, is deemed dispensable from a set of explanations. It is now time to delve in greater detail into the nature of this practice. How exactly do we con-struct a black box? How do we determine what to include and what to leave out? How does one judge whether the construction has achieved its intended purpose?

The following chapters aim to provide a more general, precise, and sys-tematic philosophical analysis of the practice underlying all the examples described in the previous pages. A precise definition of black boxes will have to wait until Chapter 5. First, we shall explore the foundations of an impor-tant scientific construct: the concept of *placeholder*.

4

Placeholders

Mihi a docto doctore
Domandatur causam et rationem quare
Opium facit dormire.
A quoi respondeo,
Quia est in eo
Vertus dormitiva,
Cujus eat natura
Sensus assoupire.[*]

—Molière, *Le Malade imaginaire*

§4.1. Introduction

Where do we stand on our conceptual map? The main goal of this book is to present and examine a form of explanation called "black-boxing." Black boxes, I contend, promise to reconcile the main insights of both reductionism and antireductionism, thereby revealing the productive role of ignorance in science. The historical excursus in Chapter 3 played a dual role. First, it showed that the construction of black boxes is widespread across the sciences. Second, it emphasized how this nuanced and complex practice may accomplish a variety of functions. Darwin, Mendel, and the subsequent biological syntheses illustrate how black boxes effectively set aside questions and problems that, for the time being, are intractable. Skinner and his fellow radical behaviorists constructed black boxes to expunge entities, such as mental states, that were deemed unsuitable for scientific investigation, in spite of their hallowed place in traditional nineteenth-century psychology. Modern economics tells a different story. Friedman and other neoclassical

---

[*] "The learned doctor asks me // The cause and reason why // Opium sends to sleep. To him I make reply // Because there is in it // A virtue dormitive // The nature of which is // The senses to allay." Translated by Sir William Maddock Bayliss.

Black Boxes. Marco J. Nathan, Oxford University Press. © Oxford University Press 2021.
DOI: 10.1093/oso/9780190095482.003.0004

economists treated feelings, preferences, and other subjective values as an inappropriate subject matter for their discipline. The rejection of mentalism here becomes a fruitful strategy to draw boundaries, fine-tune the principles of a particular field, and screen off rival methodologies. Setting these and other significant differences to the side, all these endeavors are accomplished by the same overarching strategy: the construction of black boxes.

Now that the stage is set, it is time to address more directly some issues that have been lurking in the background. What is a black box? Under what conditions is the introduction of a black box appropriate? How do these constructs accomplish such a diverse set of tasks? Answering these questions will involve a two-tiered approach. The present chapter provides some foundations. The following Chapter 5 will apply these notions and mold them into a general analysis, leading to an overarching definition.

Our starting point is a basic observation. The four historical illustrations make clear that, at the most general level, black boxes are placeholders. They are coarse-grained representations—pictorial depictions or verbal descriptions—that stand in for entities whose nature and structure are either unknown or deliberately set to the side. But what exactly are placeholders? How are they introduced? What role do they play in science? This chapter sharpens these questions and sketches some answers.

To keep the discussion focused, section 4.2 kicks off by introducing a concrete example: biological fitness. Specifically, I present two commonplace theses. First, fitness is commonly defined as a dispositional property, namely, the propensity of an organism or trait to survive in a specific environment. Second, fitness attributions are coarse-grained characterizations of biological properties that can be redescribed, more accurately and perspicuously, at finer scales. Intuitive as they are, these two tenets are jointly in tension. Section 4.3 spells out my argument, revealing its generality, and sketches a way out of the impasse. The problem, I argue, is independent of fitness per se. It affects any dispositional property that supervenes on a lower-level basis. My solution, simply put, involves noting an ambiguity in the notion of placeholder, which may replace two very different sets of entities. Section 4.4 clarifies my proposal with an array of illustrations, from mundane disposition ascriptions to scientific posits. Section 4.5 wraps up the discussion by systematizing my distinction between two types of placeholders. *Frames* stand in for a range of behaviors in need of explanation. *Difference-makers* stand in for the mechanisms, broadly construed, that produce the behaviors in question.

Before getting on with it, a quick note to readers. We are now embarking on a foundational analysis of black-boxing that will keep us busy until the end of Chapter 5. Empirically oriented scholars with little interest in philosophical discussions should consider glancing over the next two sections and focus on the illustrations and systematization in the second half of the chapter. Still, I should stress that abstract discussions of dispositional properties and explanations are crucial to fully comprehend the role of placeholders in science.

§4.2. Two Theses about Fitness

Here is a motivating example, to get us started. My analysis of placeholders begins with a short detour into contemporary biology. Fitness is a core ingredient of Darwin's framework and has remained a cornerstone of the theory of evolution ever since. As we saw in section 3.2 of Chapter 3, for a population to evolve by natural selection, it must contain heritable variation in fitness. The reason is simple. Selection fuels evolution in two different ways. It may affect an organism's *viability*—its ability to survive—or its *fertility*, that is, its capacity to reproduce. When members of a population differ along either (or both) dimensions, and such differences are heritable, at least in part, that population can evolve by natural selection. But what exactly is fitness? And what role does fitness play in evolutionary explanations? I now take up both questions, in turn.

Can we provide a precise definition of fitness? An intuitive way to approach this task is to ask: how does one determine the relative adaptedness of organisms or traits in a population? Simply put, there are two main options. One is to observe and compare actual frequencies. If individuals bearing trait *a* survive to adulthood more often or typically leave behind more offspring than organisms bearing trait *b*, this is evidence that *a* is fitter than *b*.[1] Alternatively, one can examine the physical constitution of the organisms under scrutiny. Occasionally, this merely requires educated common sense. Noting that rabbits are preyed on by foxes, which are excellent runners, suggests the (accurate) conjecture that speed and stamina contribute to the

---

[1] Strictly speaking, talking about the "fitness" of *x* is ambiguous, as it may refer either to the fitness of *trait x* or to the fitness of *organisms* bearing trait *x*. Nevertheless, for the sake of brevity, I shall often make unqualified references to the "fitness of *x*," leaving it up to the reader to determine the appropriate interpretation, depending on the context.

fitness of both species. In other cases, determining which anatomical characteristics will make organisms "fitter" necessitates more sophisticated forms of theorizing, such as mathematical modeling.

Discerning frequency patterns and studying physical makeup are effective ways of measuring fitness, which provide useful proxies. Yet, a little reflection shows that neither strategy constitutes an adequate definition of fitness. In order to see why this is so, it is instructive to address the limitations of both proposals.[2]

The futility of any general definition of fitness as a function of traits should be fairly obvious. The fittest *Bacteriophage* λ, the fittest *Drosophila melanogaster*, and the fittest *Homo sapiens* share no physical characteristic that sets them apart from less-fit conspecifics. Survival and reproductive success are heavily context-dependent. Different adaptations will be beneficial across organisms and, even when we focus on a single species, one and the same trait may be advantageous in some environments and detrimental in others. In short, what all fittest organisms have in common is surely not any feature of their physical makeup. What about the alternative route?

The shortcomings of defining fitness based on actual patterns of survival and reproduction are subtler. One problem is that this encourages the charges of circularity raised by foes of evolution at least since Spencer coined the—somewhat misleading—expression "survival of the fittest." The issue here is that natural selection is often presented by the slogan that organisms who survive are fitter. One can then legitimately ask: which individuals are the fitter ones? If the reply is that fitter organisms are those who survive longer and leave more offspring behind, we get stuck in a vicious circle. From this standpoint, it cannot fail to be the case that the organisms presently on the planet are the fitter ones, turning the principle of natural selection into an analytic truth, as opposed to an empirical discovery.

A more troubling difficulty with the analysis presently under consideration pertains directly to biological practice. Identifying fitness with actual patterns of survival and reproduction fails to distinguish the consequences of bona fide differences in viability and fertility from the outcome of spurious accidents, like genetic drift, or even brute luck. To illustrate, assume that *p*-organisms are better adapted to their environment than *q*-ones. For instance, they are stronger, faster, energetically more efficient, and better at avoiding

---

[2] The philosophical overview of fitness throughout this section draws from influential theoretical discussions, such as Godfrey-Smith (2014) and, especially, Sober (1984, 2000).

predation. For this reason, p-variants are more prevalent, in the envisioned ecosystem, than q-types. Still, as a result of a wildfire, a large portion of the p-population is wiped out. In this hypothetical scenario, more q-organisms survive than p ones, and will therefore leave behind more offspring, at least for a few generations. Yet, this is due to casualty and not—as the definition of fitness as actual survival and reproduction patterns implies—due to differences in fitness between p-types and q-types.

The extent to which these theoretical considerations affect scientific practice is debatable. Working biologists often employ actual frequency patterns as a proxy for fitness, and this seems to work reasonably well. Is there really a pressing need for change? At the field-work level, the answer may well be negative. Things are different, however, from a pedagogical perspective. Here, more accurate and perspicuous definitions of biological fitness, which overcome the conceptual limitations just rehearsed, contribute to making the content of evolutionary theory crisper and clearer.

So, what is fitness? Can we find more adequate definitions? Various proposals have been developed in the literature. An influential one is the "propensity interpretation of fitness."[3] On this view, fitness refers neither to the actual number of offspring spawned by an individual or trait relative to a reference point, nor to the physical constitution of organisms. Moreover, it is not assumed as a primitive, undefined term of evolutionary theory. Rather, in the words of its main proponents, "fitness can be regarded as a complex *dispositional* property of organisms. Roughly speaking, the fitness of an organism is its propensity to survive and reproduce in a particularly specified environment and population" (Mills and Beatty 1979, p. 270).

The propensity interpretation is not devoid of controversy. For one, providing precise mathematical characterizations of an organism or trait's expected number of offspring is no trivial task. Furthermore, ascribing an expected number of offspring to members of a population is not necessarily a reliable predictor of evolutionary change. Nonetheless, let us ignore these difficulties. Mills and Beatty's dispositional interpretation of fitness does mitigate the problematic implications mentioned earlier. Readers who find this analysis misguided are advised to skip to other examples in the following. As we shall see, fitness is just an instance of a much broader phenomenon.

With all of this in mind, set definitions of fitness aside and focus on the second issue raised at the outset. What role does fitness play in evolutionary

---

[3] For a rival approach, which I shall not discuss here, see Rosenberg (1983).

theory? How does it contribute to biological explanations? To make these questions more concrete, consider an example originally developed by Sober. Envision a hypothetical scenario where, over the course of a year, two chromosome inversions change frequency in a population of *Drosophila melanogaster*, as a result of selection pressures. What could be going on? The variation at the population level can be explained, at least provisionally, by positing that one type has greater fitness than the other. This, however, is hardly the end of the story. As researchers further inquire into the physical basis of these differences in frequency, they might discover that the chromosome in question produces a thicker thorax, which enhances the fly's insulation, making it more resistant to rigid temperatures. We now have two different accounts of the evolution of the *Drosophila* population. The higher-level depiction explains the change in frequency of the two chromosome inversions by appealing to differences in fitness between the two types. The lower-level characterization captures the evolutionary trajectory by describing the physiological effects of the chromosome inversion. What is the relation between these two explanations? Sober suggests:

> Once [a detailed] physical characterization is obtained, we no longer need to use the word "fitness" to explain why the traits changed frequency. The fitness concept provided our initial explanation, but the physical details provide a deeper one. This does not mean that the first account was entirely unexplanatory. Fitness is not the empty idea of a dormitive virtue. The point is that although fitness is explanatory, it seems to be a placeholder for a deeper account that dispenses with the concept of fitness. (Sober 2000, pp. 75–76)

What are we to make of this? Intuitively, there is an asymmetry between an organism's fitness and the underlying physical properties. (Here, the expression "physical property" should be interpreted quite generally, encompassing not only the ontogenetic characteristics intrinsic of organisms, but also the phylogenetic, environmental, and other relational conditions that are relevant to the individual in question.) The asymmetry can now be stated along the following lines. Physical properties, thus broadly construed, determine how fit organisms are, in the sense that two individuals who are physically identical must also be identical with respect to their fitness value. In contrast, the converse relation typically does not hold. The viability or fertility of an organism does not constrain its constitution. Individuals who share the same expected life span or an identical propensity to beget a particular number of offspring need not, and usually do not, share all their ontogenetic and environmental properties. We can express this asymmetry in contemporary philosophical jargon by claiming that the fitness of an organism *supervenes* on its physical base.

Let us return to our original question. What is the relation between "higher" and "lower" explanatory levels? Sober provides an effective metaphor when he characterizes fitness as a placeholder. Fitness stands in for ontogenetic, phylogenetic, and environmental processes that are responsible for producing the appropriate kind of variation. In other words, talking about fitness is a convenient way to refer to the physical properties that underlie the viability and fertility of organisms. As noted in our discussion of Darwin and Mendel, this can be done regardless of whether the detailed structure of the underlying mechanisms is known and can be stated explicitly. Once the nature of the underlying mechanisms is revealed, there remains an—open—question of whether the placeholder should be discarded or retained.

Time to tie up some loose ends. The present section introduced two widespread and relatively uncontroversial claims regarding biological fitness. First, fitness is commonly defined as a dispositional property, namely, the propensity of an organism or trait to survive and reproduce in a particular environment. The second tenet states that, since fitness supervenes on the underlying physical base, it is a placeholder for a deeper analysis that dispenses with the concept of fitness altogether. So far so good.

Here is the rub. Intuitive as they may seem, when considered separately, these two theses, taken together, are in tension. Once we combine them, they generate an apparent dilemma. *Qua* placeholder, fitness is explanatory. It provides a rough causal explanation of relevant differences in viability and fertility. In contrast, *qua* disposition, fitness does not convey any causal knowledge. This is because dispositions explicate the underlying pattern of behavior without causally explaining it.[4] Section 4.3 lays out this argument in full. I begin by clarifying the tension and showing that it is perfectly general, affecting all dispositional or supervenient properties. Next, I suggest a solution that dispels the illusion of conflict and its paradoxical flavor.

---

[4] Simply put, an *explanation* provides causal information, broadly construed, about *how* or *why* an event, or pattern of events, occurs. An *explication*, in contrast, spells out a semantic analysis, a definition of the concept in question.

§4.3. Are Dispositions Explanatory?

Cashing out the argument presented at the end of the previous section requires us to delve into the nature of a class of properties that has received much attention in philosophy: *dispositions*. While, so far, I have appealed to an intuitive understanding of dispositions, a succinct definition will make the following discussion more rigorous. Dispositional properties express the capacity, the ability, or the tendency of entities to act, under particular circumstances. For instance, solubility captures how salt dissolves in water and fragility underlies the tendency of glass to shatter when struck with force. Mills and Beatty's propensity analysis defines fitness along the same lines. The fitness of organisms and traits, on their view, is a propensity to survive and reproduce, relative to specific environments. This section discusses whether and how dispositions contribute to scientific explanations.

It is common to presuppose, more or less explicitly, that dispositions causally explain the behavior of entities that satisfy them. Indeed, many philosophers would consider this a truism to be taken for granted.[5] Is the solubility of salt not the reason that salt dissolves in water? Does a dropped vase not shatter because it is fragile? From this perspective, it is hardly surprising that, although Sober treats fitness as a placeholder standing in for, and replaceable by, a deeper analysis, he distinguishes it from the idea of a dormitive virtue, that is, an empty, question-begging circular analysis.

Things, however, are not quite that simple. How exactly does a disposition account for the behavior of the entities to which it is ascribed? What kind of knowledge does it provide, and how does it convey it? On what basis should the explanatory power of solubility, fragility, and fitness be distinguished from the empty idea of *virtus dormitiva*? For the sake of simplicity, let us begin by focusing on mundane examples: salt, glass, and the like. In due time, I shall generalize the argument to scientific dispositions like fitness.

In his celebrated masterpiece, *The Imaginary Invalid*, the French playwright Molière mocks a group of physicians who purport to uncover the sleep-inducing quality of opium by ascribing to it a *virtus dormitiva*. What triggers the humorous effect in this goofy attempt? The issue cannot be falsity. Opium, after all, does have the power to put people to sleep. The problem is rather vacuity, complete lack of informativeness. Saying that substance

---

[5] The reason, I argue elsewhere, is an assumption typically presupposed without further motivation, namely, that dispositions are properties of entities (Nathan 2015a).

x has a dormitive virtue cannot explain why *x* puts people to sleep because ascribing a dormitive virtue to *x* is just a fancier way of saying that *x* induces sleep. The entailment is analytic—or quasi-analytic[6]—turning the alleged explanation into a restatement. Molière skillfully plays with this intuitive idea that paraphrases are explicative, not explanatory.

None of this is especially novel or controversial. The aspect of Molière's insight that I want to stress is how broadly applicable it is. Having a dormitive virtue is being able to induce sleep. Thus, *virtus dormitiva* cannot explain sleep-induction because having a *virtus dormitiva* is synonymous with having the power to induce sleep. This being so, is an ascription of solubility to *x* not to say that *x* has the capacity to dissolve? Is the meaning of fragility not expressible as a tendency to shatter when struck? Answering in the positive entails that solubility cannot account for the dissolving of salt and shattering is not causally explained by fragility. Setting aside the wit—or lack thereof—these examples are structurally analogous to the mockery staged by Molière in *The Imaginary Invalid*. Counterintuitive as it may sound, it appears that none of these dispositions is explanatory after all, at least, not on any causal reading of explanation.

What about scientific dispositions such as biological fitness? Recall that, on the propensity interpretation, stating that *a*-flies are "fitter" than *b*-flies means that the former have a higher propensity to survive and reproduce than the latter. Then, can fitness attributions causally explain the distribution of traits in the population of *Drosophila*? Is fitness any more informative

---

[6] The unorthodox expression "quasi-analytic" calls for elucidation, which will take us on a slight detour into analytic philosophy. Readers with little interest for such technicalities are advised to skip this footnote entirely. In the central decades of the twentieth century a close connection was noted between dispositions and subjunctive conditionals. Thus, the fragility of a glass entails that said object would shatter, if struck. This observation was generalized into a "simple conditional analysis" of dispositions (SCA): *x* is disposed to *D* when *C* iff *x* would *D* if it were the case that *C*. Basic as it may sound, this thesis was endorsed by eminent philosophers, including Ryle, Goodman, Quine, and Mackie. However, there is now a widespread consensus that this analysis is fatally flawed because the connection between disposition and entailed conditional breaks down in cases of "finkish" dispositions, where objects temporarily lose or acquire dispositions, or when the manifestation of a disposition is "masked" or "mimicked." The philosophical community is still divided on what should replace the SCA. Some responded by replacing the "simple" conditional with a more sophisticated one. Others, such as myself, attempted to salvage the SCA (Nathan 2015a). Some explored the prospects of non-conditional analyses. Others abandoned altogether the search for explication in favor of a non- reductive account of dispositions. An assessment of these routes lies beyond the scope of this work. The point is that most authors accept the existence of a connection between dispositions and subjunctives. How is this connection to be cashed out? Advocates of conditional analyses, in the original or revised forms, will presumably maintain that the relation between dispositions and associated behaviors is analytic. Naysayers will likely resist identifying such connection with full-fledged analyticity. What is it then? For lack of a better term, I call this weaker form of entailment "quasi-analytic."

than solubility or fragility? Is any disposition at all different from a *virtus dormitiva*? It seems hard to escape a negative answer to all these questions.

We can now recast the puzzle raised at the end of section 4.2 in the form of a general dilemma. We began with two theses about fitness, widely accepted both in biology and in philosophy. First, since the fitness of organisms and traits supervenes on its physical properties, fitness is a placeholder that stands in for a more detailed, lower-level causal explanation. Second, fitness is a dispositional property, as prescribed by the propensity interpretation. In the first part of this section, we saw that these two tenets lead to opposite conclusions. The argument is simple. If fitness is a placeholder standing in for the underlying mechanisms that produce the relevant variation in populations, then it becomes clear how it can partake in causal explanations. Fitness is an abstract characterization of the causal factors that produce variation, where the nature and structure of these mechanisms are left unspecified. At the same time, conceived as a disposition, fitness cannot explain. This is because, as emphasized by Molière, dispositions explicate, but do not causally explain their associated behavior. This implies that fitness is both explanatory and non-explanatory, an overt contradiction. Something must have gone wrong. What could it be? To answer this question, it might help to visualize the argument in a schematic form:

(i) Fitness is a propensity, a disposition of organisms (Mills and Beatty).
(ii) Dispositions are explicative, but not causally explanatory (Molière).
(iii) Fitness is a placeholder for mechanisms (supervenience).
(iv) Unlike dormitive virtues, placeholders are causally explanatory (Sober).
(v) ∴ Fitness is both explanatory and non-explanatory (⊥).

The remainder of this section dispels the apparent paradox. I will stick to common wisdom and argue that both theses are ultimately correct. *Qua* placeholder, fitness is perfectly explanatory, in a causal sense. But fitness can also be analyzed as a dispositional property and, as such, it provides no causal explanation. Reconciling these two prima facie incompatible claims will require disambiguating the notion of a placeholder. I will do so by drawing a distinction between two kinds of placeholders in scientific explanations.

Before doing so, let us explore a different route. Attentive readers will have surely noted that, to avoid the contradiction (v), it is sufficient to reject one of the previous four statements. Could any of these assumptions be dropped? To answer this question, we must gauge our alternatives.

The first option is to drop assumption (i), which is tantamount to denying that fitness expresses a dispositional property. At first blush, this might look like a painless opt-out. Have I not myself acknowledged that the propensity interpretation of fitness has counterintuitive implications and is not universally embraced? Yet, upon further reflection, this solution turns out to be a non-starter. True, biologists often define fitness in non-dispositional terms. As noted in section 4.2, actual frequency patterns of survival and reproduction are often used as handy proxies. Still, this hardly addresses the problem at hand, namely, the explanatory power of fitness attributions. Actually, if anything, it exacerbates it. Allow me to explain. The tension we are trying to dispel is generated by the observation that, if fitness is defined as a propensity to survive and reproduce, then one cannot explain why a is fitter than b by noting that a is fitter than b. This being so, defining fitter traits as those that are prevalent in a population is surely not to make the claim that a is fitter than b any more explanatory. *Au contraire*, it makes it fully analytic and, therefore, *less* explanatory from a causal perspective. In short, so far, no one has been able to provide a definition that makes fitness causally explanatory, without identifying fitness with propensities, actual frequencies, or specific traits. To make things worse, the problem is does not pertain to fitness per se. The preceding dilemma can be generated whenever we have a supervenient property, many of which are bound to be dispositional in nature. This puts us right back where we started.

These considerations motivate a more promising strategy: drop assumption (ii) and concede that dispositions are indeed causally explanatory. I have no knock-down refutation of this route—KO punches do not come easy, in either science or philosophy. Indeed, I will ultimately embrace a solution that attributes a specific explanatory value to dispositions. Still, for the time being, let me stress that allowing disposition ascriptions to causally explain the behavior associated with them (quasi) analytically is a hard bullet to bite. Now, many readers will have little trouble accepting that fitness explains variation, that fragility explains shattering, or that solubility explains the dissolution. But, then, why does dormitive virtue not explain the sedative power of opium? Critics will retort that the comparison between fitness and *virtus dormitiva* is simply preposterous. Fitness captures an important property of biological populations, a constitutive ingredient of natural selection and evolutionary explanations. Dormitive virtue, in contrast, is a made-up term coined by a witty playwright to mock pompous scholasticism in medicine. The two are not on a par. I agree. Still, on what grounds should we draw this line? What distinguishes natural kinds from cooked-up impostors? Without a principled criterion, one cannot justify the claim that some dispositions are explanatory while others are merely explicative.

What about sacrificing (iii)? Is fitness really a placeholder that stands in for variation-producing mechanisms? Some readers might quibble with— or flat out reject—the specifics of my own account of placeholders sketched in the sections and chapters to follow. Yet, *that* fitness is a placeholder is harder to deny. The reason is simple. Most, arguably all, biological properties like fitness supervene on a lower-level basis, which, as noted, means that any change in fitness entails a corresponding change in physical constitution, but not vice versa. Hence, talk about fitness can always be replaced with richer, more precise characterizations of underlying physical features. From this standpoint, it is hard to avoid the conclusion that fitness ascriptions are placeholders standing in for deeper, micro-depictions.

This leaves premise (iv). The final strategy to avoid the—allegedly— contradictory conclusion is to accept that properties such as fitness are placeholders, but to deny that placeholders are causally explanatory. This gets us out of the frying pan and into the fire. Even if one could successfully show that fitness does not explain variation, the general problem is far from solved. Basically all properties that supervene on a lower-level basis can be treated as placeholders. Is no supervenient property explanatory? Are we really willing to bite this bullet and deny that biology, neuroscience, psychology, economics, and all other special sciences provide any explanation, unless they posit non-supervenient properties? What would such properties be? This radical and unpalatable form of fundamentalism flies right in the face of centuries of explanatory successes of the special sciences.

In sum, all four assumptions appear sound. Rejecting any one of them comes at a high cost. At the same time, these statements jointly imply that supervenient properties, such as fitness, are both explanatory and non-explanatory, an overt contradiction. How should we proceed, then?

My proposal boils down to a "Hempel-Davidson-style" dispelling of the appearance of the paradox.[7] Upon further scrutiny, there is no contradiction; all four premises are correct. Specifically, I defend the tenability of

---

[7] This is a nod to Hempel (1945) and Davidson (1970), who present apparent paradoxes and purport to resolve them by reconciling seemingly incompatible assumptions.

two propositions. First, dispositions are explicative but not causally explanatory. Dispositions are like dormitive virtues: they capture behaviors without explaining them. Second, dispositions are higher-level properties that stand in for unspecified mechanisms without being identical to them. As such, I shall argue, they may provide bona fide causal explanations.

My strategy to enact this reconciliation and dispel the paradoxical flavor is to diagnose an ambiguity in the notion of placeholder. Dispositions like "fitness" and other expressions denoting supervenient properties may stand in for two different sets of entities. On the one hand, they may take the place of mechanisms that produce specific patterns of behavior. When this is the case, the placeholder functions as a *difference-maker*. On the other hand, higher-level properties may stand in for the target range of behaviors that one is attempting to explain, in which case the placeholder functions as a *frame*. It explicates and lays out these behaviors but does not causally explain them.

Section 4.4 illustrates this frame vs. difference-maker distinction with examples that should by familiar to most readers. Next, section 4.5 provides a more systematic characterization of these two types of placeholders, which will provide the key for our general analysis of black boxes in Chapter 5.

Before moving on, let me stress again that expressions such as "mechanism" and "behavior" should be interpreted in the broadest and most ecumenical sense. In the present context they are essentially blanket terms that designate objects, processes, structures, physical interactions, and many other kinds of entities and activities that function either as explanantia or explananda in science. Also, as noted in Chapter 1, the idea of a property or description being "higher vs. lower level" or "macro vs. micro" is always relative to a choice of level of explanation.

§4.4. Placeholders in Scientific Theory and Practice

The previous section argued that macro-properties supervening on micro-bases may function as placeholders in two different ways. *Qua* "frames" they stand in for patterns of behaviors to be explained. *Qua* "difference-makers" they take the place of mechanisms that produce these patterns. Frames are explicative. Difference-makers are causally explanatory. This section illustrates the distinction with examples, beginning with a familiar dispositional property.

§4.4.1. Solubility

Imagine your five-year-old niece querying you about what happened to that handful of salt crystals that you just poured into a boiling pot of water. You reply by promptly noting that salt is soluble and, as such, it dissolves in water. What kind of question did she ask? And what sort of answer did you provide? Note that the issue, as just posed, is ambiguous, as your young interlocutor could be asking either of two questions. On the one hand, she could be inquiring into the behavior of salt. On the other, she could be wondering about what produces the behavior in question. Accordingly, your appeal to solubility provides two different, but equally legitimate types of answers.

First, suppose that your niece is vaguely familiar with salt. She knows what salt is. She has seen salt crystals in shakers and tasted salt in food. Yet, she does not know much about its behavior. What happens when one immerses salt in water? Does it float? Does it explode? Does it become invisible? Does anything happen at all? Appealing to solubility effectively answers this question: salt dissolves in water, a piece of information which could also be obtained by observing salt in water. Here, the ascription of solubility to salt is perfectly informative without being causally explanatory in the least. Solubility indicates the relevant behavior, *what* happens to salt, without saying anything at all about *why* or *how* the behavior itself occurs.

Next, consider a variant of this scenario where your young niece has an altogether different question in mind. She has learned in school that certain substances, such as salt and sugar, dissolve in water. She has also been shown that not all substances are water-soluble. Glass, sand, and plastic, for instance, are not. Intrigued by these remarks, she would like to understand this difference in behavior. Note that both her question and your answer may be phrased in exactly in the same terms. "What happened to salt in water? Salt is water-soluble: it dissolved." Yet, the implicit contrast class is completely different. In the former case, she wants to know that salt dissolves in water, as opposed to floating, exploding, changing color, remaining unaltered, etc. In this latter case, she wonders in virtue of what salt and sugar typically dissolve in water, whereas sand, glass, and plastic do not. Hence, pointing out *that* salt dissolves will not cut it. We need to say something about *why* or *how* it does.

A complete description of the sophisticated biochemistry of salt would be inappropriate for a child. Yet, pointing out that salt is soluble, that is, it has mechanisms that make it dissolve in water, is a much simpler, but still informative explanation.

In short, the same statement—the ascription of "solubility" to salt—may provide an appropriate answer to two kinds of inquiry. First, it may specify that salt dissolves in water. Second, it may explain how or why salt behaves the way it does. In both cases, solubility is a placeholder. But it is a placeholder of a very different kind. In the former case, "solubility" stands in for a pattern of behavior: it describes salt as dissolving in water. In the latter case, "solubility" stands in for the physical or chemical processes triggering the production of said behavior: the mechanisms that make salt dissolve in water. Let us apply the terminology introduced at the end of the previous section. When solubility stands in for patterns of behavior, it functions as a *frame*. When it stands in for mechanisms, it is a *difference-maker*, which provides a higher-level causal explanation of the behavior in question.

Some readers might feel inclined to dismiss the use of solubility *qua* frame as having marginal significance. This would be a mistake, fueled by the pernicious tendency among contemporary philosophers to privilege explanation as the paramount scientific goal, at the expense of all others.[8] Neither kind of placeholder is primary. Sometimes, an inquiry is advanced by refining the range of behaviors under scrutiny. Other times, what is sought is an explanation of these patterns. Explanation is an important scientific goal. Yet, it is hardly the only, or the most prominent one.

These considerations will be further developed in the ensuing chapters. In the meantime, let us elucidate our distinction between placeholders—frames vs. difference-makers—with some actual scientific examples.

§4.4.2. Fitness

We are now in a position to address the puzzle introduced in section 4.2. The stage was set by outlining two widely accepted theses. First, fitness is typically defined as a dispositional property, namely, a propensity of organisms

---

[8] Interestingly, prominent philosophers of science, such as Hempel (1965), explicitly warned against the perils of this asymmetry. The disproportional attention currently devoted to explanation at the expense of prediction, description, and other scientific aims witnesses the extent to which Hempel's warning has been left unheeded.

or traits to survive and reproduce in specific environments. Second, as Sober aptly puts it, the uncontroversial observation that the fitness of an organism or trait supervenes on its physical basis suggests that "although fitness is explanatory, it seems to be a placeholder for a deeper account that dispenses with the concept of fitness" (2000, p. 76). We noted how these two theses are in tension. Prima facie, they imply that macro-properties such as fitness are both explanatory and non-explanatory. I suggested that the contradiction can be avoided by disambiguating two senses in which fitness is a placeholder. On the one hand, it may stand in for patterns of variation within a population, in which case it is explicative but not causally explanatory. On the other hand, it may stand in for the mechanisms that produce the relevant variation within a population, making appeals to fitness explanatory. It is now time to elucidate the proposal in greater detail.

To make the discussion more concrete, recall Sober's hypothetical scenario, where two chromosome inversions change frequency in a population of fruit flies over the course of a year, as a result of selection pressures. Here, the statement "a is fitter than b" may address two different inquiries.

First, one may be looking into which of the two genotypes, a or b, is fitter. In this case, the statement "a is fitter than b" can be read as claiming that flies with genotype a have a tendency to survive longer and to beget more offspring than flies with genotype b. Thus understood, the difference in fitness between the two genotypes does not explain how or why these distribution patterns are observed, any more than *virtus dormitiva* explains the sedative power of opium. Yet, this hardly makes such ascriptions trivial, vacuous, or uninformative. On the contrary, pointing out that a-type flies are fitter than b-type ones makes an important contribution. It is an effective way to set up the explanandum, that is, to lay out the observed patterns of variation in the population of fruit flies, which is in need of explanation. Still, this explication, in and of itself, lacks causal explanatory force.

The statement "a is fitter than b" may also address a second kind of inquiry. Suppose that, a year into the study, the number of type-a flies has increased dramatically, whereas type-b flies significantly decreased—literally dropping like flies. One can explain this observation by pointing out a difference in fitness between a and b. Once again, fitness acts as a placeholder. Yet, here, it cannot stand in for patterns of variation. This would make the analysis uninformative, as these patterns are already presupposed in the question. Fitness now stands in for the processes responsible for producing the variation in question. In our example, these are the mechanisms, broadly construed, that allow flies with genotype *a* to produce a thicker thorax, providing better insulation during colder winter months.

In sum, fitness can be defined dispositionally, as the propensity of organisms and traits to survive and reproduce. Thus construed, fitness may function as two distinct kinds of placeholders. First, it may stand in for distribution patterns. This lays out the explanandum by stating *that* traits vary and by how much, without addressing *why* they do. Second, fitness can also act as a placeholder standing in for the properties that underlie the production of the traits in question. As noted by Sober, this shallower explanation can be eliminated and replaced by a deeper account, which dispenses with the concept of fitness altogether. The bottom line, once again, is that there are two kinds of placeholders: frames and difference-makers. As we shall see in Chapter 6, keeping them distinct is the key to understanding the role of fitness in biological explanations, from Darwin to our day and age.

§4.4.3. Psychological States

Both examples so far involve either a textbook disposition (solubility) or a concept that can be analyzed dispositionally (fitness). What about macro-properties that do not obviously fall in either category? The distinction between kinds of placeholders—frames vs. difference-makers—applies to supervenient properties of all sorts, regardless of whether or not they are dispositional. To establish this, I return to our discussion of psychological states.

What role do mental states play in psychology? Since the demise of behaviorism, many philosophers and theoretically minded psychologists have rejoined forces with Descartes's commonsensical stance in treating mental states as the principal causes of behavior. Mary's hunger plays an important part in why she is going to the kitchen to fix herself a sandwich. And Jack's being upset with Jill is one of the reasons why he refuses to call her. Now, to be sure, various theories of mind vehemently disagree over the nature and structure of these psychological states. Functionalists, type-identity theorists, token-identity theorists, and property dualists—to mention a few influential candidates—offer very different analyses of the ontology and epistemology of mental states. Yet, contemporary psychology and philosophy of mind agree that, contrary to the strictures of behaviorism discussed in section 3.4 of Chapter 3, mental states contribute to triggering human conduct.

I have no quibbles with any of this. Still, I do want to emphasize that this is only part of the story. Mental states play two distinct and equally important roles within psychological investigations. On the one hand, as just noted, mental states may function as *difference-makers* in the production of behavior. On the other hand, mental states may be *frames* standing in for patterns of behaviors to be explained. Allow me to clarify.

Imagine that Taylor and Dana spend a lot of time together. They make dinner plans. They go out dancing. They talk on the phone daily. And they often blush in the presence of one another. All these behaviors—which note, constitute a wildly heterogeneous bunch—can be explained, at least provisionally, by positing that Taylor and Dana are *in love* with each other. Could one provide lower-level explanations of these phenomena? Of course! For instance, we could provide a detailed account of how, in the presence of Taylor, Dana's body releases adrenaline, speeding up heart rate and dilating blood vessels, improving blood flow and oxygen delivery in facial veins. We now have two explanations of Dana's blushing: a higher-level one and a lower-level one. Does this story have a familiar ring to it? If so, it is likely because it mirrors our discussion of fitness from section 4.2.

The precise relation between these two explanations is a controversial matter. On the one hand, reductionists will argue that once a more detailed physiological explanation is obtained, there is no need to use the expression "being in love" to account for blushing. The mental state can be reduced to a deeper, lower-level analysis, thereby eliminating it. Antireductionists will retort that no such reduction can be enacted. We shall return to the reductionism vs. antireductionism debate in Chapter 10. For the time being, the issue is that both stances can be reconciled with the claim that mental states are placeholders, *difference-makers* standing in for mechanisms producing behavior. The point of contention is the nature and character of the mechanisms explaining the behavior. Reductionists will characterize them at lower levels—brain processes, environmental stimuli, and the like. Antireductionists will provide higher-level characterizations, appealing to functional descriptions, qualia, or other macro-descriptions. But the mental-states-*qua*-placeholder thesis is perfectly consistent with both stances, including radical forms of materialism and a hard-core dualism rejecting mind-body supervenience.

There is also, however, a different way in which "being in love" is a placeholder. It can function as a frame, that is, it may stand in for a range of behaviors that we are trying to explain. To illustrate, imagine that a team of psychologists sets out to investigate Taylor's psychological profile. As any ordinary human agent, Taylor engages in a vast array of behaviors, ranging from breathing, to eating, to thinking, to working, etc. How do we restrict the range of actions that our research team is going to study? An effective way to do so is to focus on the behaviors displayed by Taylor in virtue of being in love with Dana. In this sense, observing *that* Taylor is in love with Dana does not explain *why* they are out dancing tonight, any more than ascribing solubility to salt explains why salt dissolves in water rather than exploding, or why the dormitive virtue explains why opium puts people to sleep as opposed to killing them. Nevertheless, loving-*qua*-frame plays a prominent, non-explanatory role in psychological inquiries. It captures, in a clear, precise, and succinct fashion, the range of actions that we are trying to explain, that is, *how* the agent in question behaves.

In conclusion, mental states are not dispositions—at least, they are not obviously so. Yet, they are placeholders which may be used either as frames, to capture how an agent behaves, or as difference-makers to provide a higher-level causal explanation of the conduct itself. These considerations suggest a more charitable reinterpretation of behaviorism. *Pace* Ryle and (early) Wittgenstein, the manifestation of behavior, no matter how nuanced, is neither necessary nor sufficient for mental states. *Pace* Skinner, the causal explanation of behavioral patterns cannot be simply reduced to stimulus conditions supplemented by operant behavior. Mental states are higher-level causes of behavior. But this is only part of the story. And behaviorism was largely correct about the other half. The evident connection between mind and conduct allows us to use mental states as frames that capture an agent's complex behavioral patterns in need of explanation.

§4.4.4. Economic Utility

The distinction between frames and difference-makers also sheds light on the role of utility in economic theory. Suppose that, on several occasions, I observe Bill choosing apples over oranges, regardless of price fluctuations between the two items. Economists from various traditions capture this data by positing that Bill *prefers* apples to oranges in the following sense. With diminishing marginal utility, the utility Bill receives from his first unit of apples consumed is greater than the utility he receives from his first orange unit consumed. His second unit of apples will bring him less marginal utility than his first. (Whether this amount of utility is less than the utility he receives from his first unit of oranges is a separate question, that we shall set aside.) In general, it is assumed that Bill will consume a mixture of these fruits, and he will do so until $MU_x/p_x = MU_y/p_y$ for all x and y to maximize total utility obtained. Without getting into technicalities, we can simplify this by positing that Bill assigns a higher utility to apples than oranges: $u(a) > u(o)$, and these utilities can only be ranked, as opposed to quantified precisely. Note that this claim is affected by the same kind of ambiguity that characterizes all previous examples. On the one hand, the proposition that $u(a) > u(o)$ may function as a frame. In this former sense, it is a placeholder for the relevant pattern of behavior in need of explanation. On the other hand, that same proposition may act as a difference-maker. In this latter sense, it is a placeholder standing in for mechanisms that underlie and produce these observable patterns of possible and actual behaviors. How does this work?

First, imagine that the goal is to explain *why* Bill consistently prefers apples to oranges. What is going on in Bill's head when he is presented with this basic choice? The notion of utility provides a preliminary answer: Bill assigns a higher utility to apples than oranges. Note that, in the expression $u(a) > u(o)$, "u" is a placeholder. It stands in for whatever psychological mechanism underlies the choice at hand. Just as we saw in the cases of solubility, fitness, and mental states, this coarse-grained depiction can be replaced by a richer micro-explanation that makes no reference to utility, but only describes the psycho-neural processes that trigger the behavior in question. In a nutshell, in this first instance, utility is a difference-maker.

Next, consider a slightly different scenario. We embark in a systematic study of Bill's fruit-related behavior. On day one, Bill is presented with a choice between apples and oranges, and he selects the former. The following morning, confronted with a similar range of options, he still picks apples over oranges. On day three, the store is out of apples, and Bill picks pears over oranges. On day four, the store is out of oranges, and Bill selects apples over pears. It should be obvious that this depiction is somewhat simplified. Real-life agents face choice selections that are exponentially more complex and involve numerous other variables. Nonetheless, this basic scenario should suffice to establish the main point. Can we systematize all these behaviors in a coherent fashion? Utility provides a simple and convenient way of doing so. Consider the proposition that $u(a) > u(p) > u(o)$, which states that Bill assigns a higher marginal utility to apples than pears and that his utility for pears is higher than the value assigned to oranges. This simple claim allows us to make lots of predictions and retrodictions of Bill's behavior pertaining to fruit choice. Importantly, this is not a causal explanation of behavior. Yet, this frame for the behavior to be explained is an important endeavor, nonetheless. It answers the *that* but not the *how* or the *why*.

Before moving on, a historical clarification is in order. Friedman, Savage, and other early neoclassical economists who worked within the milieu of logical positivism endorsed a covering-law (C-L), or deductive-nomological (D-N) approach to explanation. On this model—which identified explanation with logical derivation from a set of laws and initial conditions, along the lines discussed in section 2.3 of Chapter 2—prediction and explanation are two sides of the same coin. If Bill's behavior can be accurately predicted from the preceding ranking of utility, then it has also thereby been explained. Contemporary philosophy of science rejects the C-L model of explanation, for well-known reasons briefly rehearsed in Chapter 5. Yet, virtually all contemporary accounts explicitly eschew the identification of prediction and explanation. As a result, they relinquish the idea that assignments of utility provide an explanation and, especially, a causal explanation of the resulting behavior.

The distinction between these two notions of utility—frame vs. difference-maker—can be used to characterize, more precisely, the relation between different approaches to economics. Mainstream neoclassical economics (NCE) treats utility, marginal or total, as an abstract mathematical construct: an ordinal ranking fully revealed by preference, with no cardinal counterpart. In contrast, much contemporary psycho-neural economics (PNE) views utility as an actual neural value, instantiated in the brain and quantitatively measurable, at least in principle. Despite sharing the name "utility," these two concepts are importantly distinct, as they presuppose different assumptions and methodological approaches. These considerations present an obvious quandary. Given these evident discrepancies, why would neoclassical and psycho-neural economists view these two notions of utility as competing alternatives?

My suggestion is simple. NCE and PNE disagree on *how* choice should be explained. Neoclassicists believe that the actual process underlying decision-making is irrelevant to economics. Psycho-neural economists consider these same mechanisms paramount, relative to their epistemic goals. Thus, utility is a different explanans across the two theories. They posit distinct difference-makers. At the same time, at a very broad level of description, NCE and PNE may converge in their use of utility as a frame, that is, on the kind of behavior that falls within the domain of economics.

On this note, what counts as "economic" behavior? How does one pick out the set of phenomena to be covered? The central aim of economics is to explore the consequences of choice. Which choices? Whenever options are characterized by differences in utility, such choices can be studied from an economic standpoint. At this general level of description, NCE and PNE converge on Robbins's definition, presented in section 3.5 of Chapter 3: "Economics is the science which studies human behaviour as a relationship between ends and scarce means which have alternative uses." Utility frames the behavior to be explained by both approaches. How are these patterns to be accounted for? Here the two theories diverge. One side opts for an idealized "as if" mathematical model. The other side cashes out a mechanistic model. In sum, the two approaches may be subsumed under the same frame, the same choice of object of explanation. But the underlying selections of difference-makers are clearly distinct.

§4.4.5. Phlogiston

All the examples discussed so far involve supervenient properties—some dispositional, others categorical—which are widely employed in everyday life, as well as in scientific theory and practice. Our final illustration draws attention to a different set of issues, by focusing on a superseded notion that has been purged from our ontology: the concept of *phlogiston*.

Phlogiston theory was introduced in 1667 by J. J. Becher. Initially developed by G. E. Stahl and his student J. H. Pott, it received its most advanced formulation toward the end of the eighteenth century in the work of Joseph Priestley and Henry Cavendish. Such theory purports to provide a systematic explanation of important chemical reactions, most prominently, the process of combustion. Simply put, the term "phlogiston" was introduced to pick out the substance, the "principle," emitted in all cases of combustion. As it turns out, there is no single substance that is emitted in all cases of combustion. Consequently, the term "phlogiston" fails to refer—although, as will become clear in the course of our discussion, this initial characterization is a crass oversimplification. These well-known observations raise some interesting problems for philosophical theories of reference and related issues concerning the advancement of scientific theories. Specifically, if "phlogiston" is non-referential, then what explains the celebrated successes of Priestley and his colleagues?

The issue of progress will be addressed, in detail, in Chapter 9. For the time being, my goal is to show that superseded scientific concepts, such as phlogiston, raise a kind of puzzle similar to the one presented in section 4.3. From the standpoint of phlogiston theory, the expressions "combustible" and "rich in phlogiston" are, for all intents and purposes, synonymous. The reason should be evident. Since phlogiston is defined as the substance that is released in the air during combustion, it follows that all and only combustible substances contain phlogiston. Consequently, phlogiston looks like solubility and fitness. While these dispositional properties explicate the underlying concepts, they appear to have all the explanatory force of a *virtus dormitiva*. This, once again, seems counterintuitive. Phlogiston is the eponymous explanatory core of phlogiston theory, not a definition. Sure, contrary to fitness or solubility, it has been discarded. But how could an analytic entailment—the stipulative identification of phlogiston-rich substances with combustible ones—turn out to be empirically false?

Attentive readers will probably anticipate my response to this quandary, which is analogous to my solution to the general puzzle. "Phlogiston" is a placeholder. And there are two types of placeholders, depending on the kinds of entities they stand in for.

To begin, recall that the chief goal of phlogiston theory is to explain the process of combustion. Combustion generates familiar patterns, where a substance is gradually transformed into another as a result of a chemical reaction. Thus, burning transforms wood into ashes, while simultaneously altering the composition of the surrounding air. Granite, in contrast, is not combustible because burning it does not produce any equivalent reaction. What role does the concept of phlogiston play in these explanations?

Here is the punchline. The statement "substance *x* is rich in phlogiston" is ambiguous, in the sense that it provides an answer to two different questions. On the one hand, it may specify the kind of behavior to be expected when substance *x* is burned, namely, that it will combust. In this sense, "rich in phlogiston" is synonymous with "combustible" in the same way that "soluble" and "tends to dissolve in water" mean exactly the same thing. Just like solubility, thus understood, cannot causally explain why salt dissolves in water, the claim that wood is rich in phlogiston cannot capture why wood combusts. The reason has nothing to do with the shortcomings of phlogiston theory per se. It is because explications are not causally explanatory. In this former sense, "phlogiston" is a frame. It picks out the class of events that the theory is supposed to account for and, as such, it is explicative without being explanatory.

On the other hand, "phlogiston" may refer to the principle, substance, or process that causes and underlies the reaction itself. The theory posits the existence of a single principle, phlogiston, that is emitted in all cases of combustion. In this second sense, "phlogiston" is employed as a difference-maker. It picks out whatever it is that makes a difference to why and how the process of combustion occurs. As such, it is perfectly explanatory. It just so happens that it provides an inaccurate, superseded causal explanation.

In sum, *qua* frame, phlogiston theory provides valid explananda; we just no longer use the terms of this old theory to describe them. *Qua* difference-maker, it provides discarded explanantia, replaced by atomic chemistry.

Incidentally, why exactly has phlogiston theory been discarded? If "rich in phlogiston" is synonymous with "combustible," and combustion is a real phenomenon in nature, then why was the concept of phlogiston eventually eliminated and replaced by atomic chemistry? The obvious reason is that talk about "phlogiston" is not perspicuous enough. In particular, it introduces a nonexistent element, phlogiston. In doing so, it fails to address a number of important distinctions. By positing a single substance that is emitted in all cases of combustion, phlogiston theory clashes into the same category of events which should be kept distinct, like the burning of wood and the heating of metals. Unsurprisingly, modern atomic chemistry fares much better on this score. By differentiating between, say, the combustion of wood and coal and the oxidation of mercury, iron, and other metals, it provides more precise causal-mechanistic explanations and picks out a more perspicuous range of behaviors. Still, these observations raise further issues. Why is the transition from phlogiston theory to atomic chemistry a *progressive* one? Addressing this and related questions requires a more systematic discussion of black boxes. We will return to it in Chapter 9. Now, it is time to wrap up our discussion of placeholders.

§4.5. Two Types of Placeholders

Time to tie up some loose ends. This chapter began by presenting the notion of biological fitness. I outlined two widespread theses. First, fitness is commonly defined as a dispositional property. It is the propensity of an organism or trait to survive and reproduce in a particular environment. Second, since fitness supervenes on its underlying physical properties, it is a placeholder for a deeper account that dispenses with the concept of fitness altogether. Plausible as they are, these two theses are in tension. *Qua* placeholder, fitness is explanatory. *Qua* disposition, it explicates but cannot causally explain the associated behavior. The rub is that dispositions are placeholders. Thus, taken at face value, our intuitive premises generate an overt contradiction. In the second part of the chapter, I suggested a way out of this impasse. My solution, simply put, involves disambiguating the notion of placeholder. On the one hand, a placeholder may stand in for the range of events to be accounted for. In this case, the placeholder functions as a *frame*: it spells out an explanandum, a behavior, or range of behaviors in need of explanation. On the other hand, a placeholder may stand in for the mechanisms, broadly construed, which bring about the patterns of behavior specified by the frame, the explanandum, regardless of how well their nature and structure are understood. When this occurs, the placeholder becomes a *difference-maker*. These types of placeholders can be visualized in Table 4.1, which distinguishes them along two orthogonal axes: their role as explananda or explanantia, and whether they stand in for mechanisms or behaviors. By drawing these simple distinctions, all four premises in the argument from section 4.3 can be maintained consistently.

We are finally in a position to resolve the general paradox introduced in section 4.3. Are appeals to solubility, fitness, mental states, and utility causally explanatory? Short answer: it depends. As frames, these constructs are not explanatory; they require explanation. As difference-makers, they are perfectly explanatory. Thus, Sober is correct in noting that appealing to fitness provides an explanation of changes in frequency patterns across populations. At the same time, his attempt to contrast fitness with *virtus dormitiva* is potentially misleading. Sure, when dormitive virtue stands in for a range of behaviors—putting people to sleep—then it hardly explains the powers of opium, or anything else. This confusion is what generates the

Table 4.1 Two Types of Placeholders in Science

Explanatory Role	Placeholder Type	Stands in for
Explanans	Difference-maker	Mechanisms
Explanandum	Frame	Behavior

comic effect skillfully crafted by Molière. Yet, when *virtus dormitiva* stands in for the mechanisms that sedate people, appeals to dormitive virtue become unproblematically explanatory. Sure, they might not be very enlightening. *Virtus dormitiva* may be closer to solubility than fitness, mental states, or utility. But a vague, general explanation is an explanation, nonetheless.

It is worth stressing that my notion of placeholder is quite broad, and intentionally so. Specifically, placeholders may range from having much to very little content. To wit, the higher-level properties discussed in this chapter—solubility, fitness, mental states, utility, phlogiston, and the like—provide a minimal specification of the structure of the mechanisms represented. Nevertheless, if all supervenient properties are placeholders, it is also possible to provide much richer, more detailed, characterizations of the underlying properties. This anticipates an important point to be developed in Chapter 5. My definition of a black box, based on the dual role of placeholders introduced in the present chapter, encompasses not only stereotypical "black boxes," but less opaque concepts, such as gray boxes, and quasi-transparent boxes too, as long as these supervene on lower, physical levels.

In conclusion, how does all of this fit into the philosophy of science? What implications does our analysis of placeholders have vis-à-vis the role and contribution of higher-level properties to scientific inquiries? Many authors have embedded these considerations into the broader context of the reductionism vs. antireductionism debate. Sober, for one, generalizes his discussion of biological fitness to the conclusion that all properties in the special sciences supervene on physical properties. This, in turn, raises an overarching question about the nature of scientific explanation. If the special sciences supervene on physics, is there a fundamental physical explanation for any phenomenon explained by other sciences? If higher level properties are placeholders that can be replaced by more detailed lower-level descriptions, are these micro-level explanations always deeper than their macro-counterparts?[9] These are precisely the type of issues that we have previously encountered in Chapters 1

---

[9] Sober explores two kinds of answers. First, he suggests that, while all non-fundamental explanations retain practical value, in principle, they are all disposable. This seems like a modest form of reductionism, since it acknowledges that, while physics is not yet able to explain everything, in principle, it could. Next, he goes on to provide a second answer, which falls in line with sophisticated antireductionism. Even if it would be possible to best explain all individual *token* events at the most fundamental level, explanations of the *type* sort cannot be paraphrased at lower levels, either in practice or in principle. For a similar argument, from the standpoint of physics, see Batterman (2002).

and 2. Once again, we are hopelessly haunted by the Homeric hazard. What is our doom? Are we going to get devoured by Scylla or drowned by Charybdis?

By now, readers will not be surprised to hear that I intend to explore a different route. There is no need to pick your poison. We can have the cake and eat it too. Specifically, reductionists are absolutely correct that it is always possible to replace a causal-mechanistic explanation of how and why a certain phenomenon occurs with a more detailed, lower-level, micro-depiction. At the same time, this does not invalidate the antireductionist tenet that many explanations in the special sciences are "autonomous," in the sense that they stand alone perfectly well, without the need to enrich them with additional micro-details. Showing how these seemingly incompatible claims can be reconciled is the ultimate task of this book. Before getting there, we have ways to go. Since my positive proposal will focus on black-boxing, the first move is to elucidate this strategy. With this target in sight, the following chapter identifies and discusses the constitutive stages of this form of explanation.

5

Black-Boxing 101

Oh, won't you
    Gimme three steps, gimme three steps, mister
    Gimme three steps toward the door?
    Gimme three steps, gimme three steps, mister
    And you'll never see me no more
                    —Lynyrd Skynyrd, "Gimme Three Steps"

§5.1. Introduction

Chapter 4 began a more systematic investigation of the black boxes first identified by our historical excursus in Chapter 3. Specifically, I introduced a distinction between two kinds of placeholders in science. *Frames* stand in for explananda, patterns of behavior in need of explanation. *Difference-makers* stand in for explanantia, mechanisms that produce and explain the patterns in question. Both kinds of placeholders will play a pivotal role in my analysis of black boxes, developed with an eye on the following issues. What is a black box and how is it constructed? How does a single concept accomplish so many tasks, across a variety of fields? Can black boxes explain the interplay between productive ignorance and knowledge?

Here is the punchline. This chapter sets out to decompose the black-boxing strategy into three constitutive phases. The first step involves sharpening the explanandum by placing the object of explanation in the appropriate context. This is typically accomplished by constructing a frame, a placeholder that stands in for the pattern(s) to be covered. For this reason, I refer to this phase as the *framing stage*. The second step, the *difference-making stage*, provides a causal explanation of the explanandum, now appropriately framed. This, simply put, involves identifying the relevant difference makers, that is, placeholders standing in for the mechanisms that produce the patterns under scrutiny. The third step, the *representation stage*, determines how these difference makers should be characterized, which features of the mechanism are to

Black Boxes. Marco J. Nathan, Oxford University Press. © Oxford University Press 2021.
DOI: 10.1093/oso/9780190095482.003.0005

be explicitly represented and which can be idealized or abstracted away. The outcome of this process is a *model* of the explanandum, a representation of the target system or relevant portions of the world.

Taken together, these three steps provide the general definition we have been looking for. A black box is a placeholder—frame or difference-maker—in a causal explanation represented in a model. The goal of the sections to follow is to elucidate this compact statement by breaking it down to its main constituents. How is an explanandum framed (§5.2)? How is it sub-sumed under a causal explanation (§5.3)? How is this causal explanation represented in a model (§5.4)? With this definition on the table, our next task, in Chapter 6, will be to apply the present analysis to our case histories, before exploring, more systematically, its philosophical implications.

Before moving on, allow me a clarification and a word of caution. Beginning with the former, I should stress from the outset that all three stages are individually necessary and jointly sufficient for the effective con-struction of a black box. Now, surely, it is not uncommon in science for one of these phases to be performed in isolation from the others. To wit, so-called exploratory experiments often complete the first two steps of the procedure described earlier, without attempting to provide a mechanistic model of the phenomena in question. Nevertheless, I maintain that the specification of a black box requires all three steps. To emphasize, there is nothing inherently wrong with research that only engages with some of these phases without bringing all three of them to completion. Still, from the present standpoint, the outcome of these inquiries falls short of a full-fledged black box. At the same time, it would be a mistake to insist that these steps must invariably be taken in any specific order. Scientific research is complex and messy, and practitioners are often pushed into a project *in medias res*. Hence, my nar-rative should not be understood as a realistic portrayal of laboratory life. It is intended as a regulative ideal, a rational reconstruction of actual empir-ical work.

Finally, here is my cautionary warning. The present chapter is admittedly dense, rich in somewhat technical analysis, and it covers a vast philosophical literature. There is a reason for this. The topics discussed here—explanatory relativity, causal explanation, and modeling—have a hallowed place in con-temporary philosophy. Yet, my rationale for inserting them in the appro-priate context transcends my scholarly due diligence. These long-lasting debates reveal why problems, commonly glossed over in much scientific lit-erature, should not be taken for granted. A proper philosophical analysis of science must pay attention to both its scientific and philosophical precursors. Having said this, frustrating my audience is neither kind nor wise on my part. Empirically minded readers with little interest in philosophical debates may want to consider glossing over the detailed discussion of the three stages of black-boxing, focus on the succinct summary in section 5.5, and dig into the real-life scientific examples and applications in Chapter 6.

§5.2.   Step One: The Framing Stage

Requests for explanation are seldom, if ever, unqualified. To wit, consider a mundane question such as: why did the match light? What is the appropriate reply? It depends. Typically, it suffices to remark that the match lit because it was struck. Still, there are situations where this response misses the mark. Imagine a legal setting where an attorney is reconstructing the chain of events that led to a gas station blowing up. It has been established that the explosion was triggered by the burning of a match. While interrogating a suspect, the attorney asks why the match lit. Replying that the match lit because it was struck would be inappropriate, teetering on contempt of court. What is in question here, presumably, is whether the match was struck deliberately, with the intention to detonate, or whether it was struck accidentally by a careless, non-malicious agent. Or think of a fictional NASA probe striking a match while examining the chemical atmosphere of a distant planet. Here, the appropriate answer to "why did the match light?" could be the sufficient quantity of breathable oxygen in the atmosphere, an observation that would be uninformative, even insulting, in the previous cases. The first scenario is more usual. The other two are admittedly far-fetched. Still, they establish an important point. Explanations come equipped, more or less explicitly, with information that provides the context in which they should be understood, processed, and assessed.

These basic remarks are hardly original. Both vintage and contemporary discussions acknowledge this relativization of explanation and related pitfalls. Before delving into the complexities of scientific framing, let us examine how instances of this phenomenon have been analyzed in philosophy.

A well-known illustration of explanatory relativity comes from a 1973 paper by Fred Dretske. Alex, after being fired, is short on cash. His friend Clyde lends him $300. Dretske notes that there are (at least) three questions that may be expressed by the string of words: "why did Clyde lend Alex

$300?" Accordingly, three distinct answers may be appropriate. First, why did Clyde lend Alex *$300*, as opposed to, say, $100 or $500? Second, why did Clyde *lend* Alex $300, as opposed to gifting him the cash? Third, why did *Clyde* lend Alex the cash, as opposed to their common friend Scrooge, who is rolling in dough? Dretske notes that the difference in stress is a pragmatic one, posing serious challenges to purely syntactic or semantic theories of explanation. This is true. But the issue remains: how should this pragmatic form of explanatory relativity be characterized and addressed?

An influential development of this pragmatic dimension of explanation is provided by van Fraassen in his 1980 book, *The Scientific Image*, a contemporary classic in the philosophy of science. The backdrop of van Fraassen's analysis is constituted by the "problem of asymmetry," the most widely debated counterexample to Hempel's "deductive-nomological" (D-N) model of explanation, common among logical positivists.[1] Imagine a tower 100 feet tall casting a shadow 75 feet long. According to the D-N model, we can explain the length of the shadow by deducing it from the height of the tower, in conjunction with assumptions on the angle of elevation of the sun, basic principles of trigonometry, and light traveling in straight lines. The problem is that these same background conditions equally allow one to deduce the height of the tower from the length of the shadow. Now, no one seriously wants to question that the tower's height explains the shadow. But does the shadow's length explain the tower? Most philosophers answer in the negative, generating a counterexample to the D-N model, which is committed to an affirmative answer.

All of this was well known by the 1980s. An explanation of the height of the tower appealing to the length of the shadow would typically be rejected. But does the shadow's length *never* explain the tower's height? Van Fraassen casts doubt on this stronger conclusion. His memorable fable, "The Tower and the Shadow," skillfully crafts a scenario where the height of the tower is, indeed, explained by the length of the shadow because of the builder's desire for the shadow to reach a specific spot at the right time of day. These considerations are molded into a pragmatic analysis, according to which an

---

[1] From this D-N perspective, also known as the "covering-law model," the explanation of an event consists in a formal derivation of an explanandum from a set of laws and initial conditions. Thus, one can explain why the match lit by observing that, given initial conditions, such as the presence of oxygen in the air, the structure of the match, and the striking of the match, together with the relevant laws of nature, the explanandum event was logically bound to happen. The problem of asymmetry was noted and presented, through a variety of equivalent examples, by various authors, especially Michael Scriven and Sylvain Bromberger.

explanation is an answer to a question Q of the form "Why P_k?" Specifically, explanatory requests are identified with ordered triplets $<P_k, X, R>$, where P_k is the "topic" of the question, $X = \{P_1 \ldots, P_k \ldots\}$ is its contrast class, and R is the relevance relation of the explanation.

Is van Fraassen sketching a pragmatic theory of explanation or a theory of the pragmatics of explanation? On the former, stronger reading—characterized as a theory purporting to fully account for scientific explanation in terms of pragmatically determined contextual features—the view is notoriously subject to trivialization worries. Still, even critics who note these shortcomings acknowledge the value of van Fraassen's contribution when understood as a theory of the pragmatics of explanation.[2] The basic insight can be paraphrased as noting the context-relativity of explanation when it comes to assessing its intended target and, consequently, its success.

Yet another notable attempt to characterize this explanatory relativity can be found in Garfinkel's 1981 monograph, *Forms of Explanation*. When the infamous bank robber Willie Sutton was in prison, he was allegedly visited by a priest, who asked him why he robbed banks. "Well," Sutton replied, "that's where the money is!" What is going on in this exchange? Simply put, the bandit and the clergyman are talking past each other. This is because they are tacitly presupposing different objects of explanation. The priest is wondering about why Sutton would choose to rob at all, as opposed to conducting an honest life. Sutton, in contrast, is addressing a different question. Given his decision to rob, why should the target be a bank, as opposed to, say, a public library or a department store? The two interlocutors have different conceptions of the relevant options. They take the same string of words—"why do you, Sutton, rob banks?"—but embed this statement in different spaces of alternatives, effectively changing the explanandum. This triggers the humorous effect. Incidentally, Garfinkel notes, this simple recipe has produced countless jokes for generations. Why do birds fly south for the winter? Because it is too far to walk!

Setting jest aside, the philosophical point—explanatory relativity—is deep and important. Explanations always take place relative to a background of alternatives, a contrast space that determines the nature of the explanandum

---

[2] As Kitcher and Salmon (1987, p. 316) put it in their scathing review of van Fraassen's theory of explanation, "Failing to appreciate that arguments are explanations [. . .] only relative to context, we assess the explanatory merits of the derivations by tacitly supposing contexts that occur in everyday life. With a little imagination, we can see that there are alternative contexts in which the argument we dismiss would count as explanatory."

and whether the explanans is ultimately successful.[3] In some cases, the salient alternatives may be stated explicitly. Why did you finish the pizza but not the broccoli? More typically, the underlying contrast class is presupposed at a tacit level. Dretske's example is a case in point. Whether the question is why Clyde lent Alex $300, why Clyde *lent* Alex $300, or why *Clyde* lent Alex $300 is unspecified. It must be inferred from the context. Similarly, Sutton's remark that banks are where the money lies is a perfectly good explanation of why Sutton robs banks, but only provided that his commitment to robbing is assumed. It should be clear from the context that the priest is not willing to take that premise for granted.

Time to take stock. All requests for explanation are subject to interpretation. Even simple, ordinary, everyday questions—Why did the match light? Why did Clyde lend Alex $300? Why does Sutton rob banks?—are tacitly equipped with assumptions, "given that" clauses, and pragmatic presuppositions that place them in the appropriate context so that they can be parsed, processed, and addressed. In other words, the object of explanation is never an event *simpliciter*. The explanandum is always an event together with a contrast class, that is, a set of alternatives $C_1 \ldots C_n$ that determine what counts as an acceptable explanation of E.[4]

---

[3] What exactly is a contrast space? Garfinkel (1981, p. 40) describes its basic structure as follows. If Q is some state of affairs, a contrast space for Q is a set of states $[Qa]$ such that (i) Q is one of the Qa; (ii) every Qa is incompatible with every other Qb; (iii) at least one element of the set must be true; and (iv) all of the Qa have a common presupposition, that is, there is a P such that, for every Qa, Qa entails P.

[4] Why must explanations thus limit our options? The answer, Garfinkel suggests, "lies in our need to have a *limited* negation, a determinate sense of what will count as the consequent's "not" happening" (1981, p. 30). This point is elucidated with the help of an example. Suppose that one morning, I go out for a drive. I am going over 110 mph when I round a bend where a truck has stalled. Unable to stop in time, I crash into the truck. Later, you chastise me: if you hadn't been speeding, you would not have crashed. That is true, I retort. But then, had I not had breakfast, I would have reached that spot before the truck. So, had I not eaten breakfast, I would not have crashed. Why do you not blame me for having breakfast? What makes my reply fishy? Here is Garfinkel's answer (1981, pp. 30–31): "My claim is based on the assertion that if something (eating breakfast) had not happened, the accident would not have happened. The problem is, what is going to count as that accident's not happening? If 'that accident' means, as it must if my statement is going to be true, 'that very accident,' that concrete particular, then *everything* about the situation is going to be necessary for it: the shirt I was wearing, the kind of truck I hit, and so forth, since if any one of them had not occurred, it would not have been *that* accident. But this is absurd, and in order to escape this absurdity and not have everything be necessary for the accident, we must recognize that the real object of explanation is not my having had that accident. [. . .] [T]he real object of explanation is an equivalence class under [a] relation [determined by a set of 'irrelevant' perturbations]. The equivalence relation determines what is going to count as the event's not happening." As Michael Strevens has brought to my attention, appealing to blameworthiness as a test for causal or explanatory relevance is problematic, as these features do not invariably go hand in hand. Causally explanatory factors like icy or oily roads may not attract blame. Vice versa, non-explainers such as negligence *sans* adverse effects can be morally relevant. Causal relevance will be discussed in detail in section 5.3. The important point, for the time being, is

Explanatory relativity is an old adage, widely noted and discussed in the philosophical literature. Nevertheless, these considerations raise some interesting theoretical issues for science, whose significance is frequently overlooked. Spelling out the background assumptions, the contrast class, of everyday explanations is daunting enough. It requires interpreting various contextual cues and parsing tacit presuppositions. And, yet, the complexity of lit matches, towers and shadows, and friends borrowing cash pales in comparison to actual scientific inquiries. Where does one even begin contextualizing explanations which presuppose evolution by natural selection, classical genetics, behavioristic psychology, or neoclassical economics? Borrowing a Firestein comment, originally directed to the constant growth of scientific knowledge, and adapting it to a slightly different context: "How does anyone even get started being a scientist?" (Firestein 2012, pp. 14–15).

To fathom the real depth of the problem, let us take a brief detour into the nature of science itself. As noted in Chapter 1, textbooks have a tendency to present theories as systematic arrays of laws and principles, obtained through an organized accumulation of data sets, and a rational discussion of their implications. This image is a drastic oversimplification, and hardly an inconsequential one. Things in actual laboratories are much messier, more chaotic, and way more interesting. Scientific fields are characterized by core problems, methods, and techniques. These rich theoretical thickets determine the range of explananda and the acceptable explanantia.

To complicate things even further, scientific fields neither provide research questions from scratch, nor do they stand alone. What look like new domains actually import assumptions, results, methodologies, and experimental techniques from predecessors and neighboring disciplines. Here is a simple albeit paradigmatic illustration. The concept of gene did not originate in molecular biology. As we saw in section 3.3, the existence of "factors," their relevance to heredity, and even their relation to chromosomes were postulated by classical geneticists who, in turn, inherited them from Mendel. And their roots arguably dig even deeper. All of this goes to show that, to spell out the relevant context of a scientific hypothesis, such as the biochemical explanation of meiosis, it is not even sufficient to provide an exhaustive characterization of cytology. Many concepts, assumptions, methodological

---

simply that explanatory relevance depends on the contrast class. This is independent of the specifics of Garfinkel's example.

standards, problems, and presuppositions have been inherited from previous paradigms or were transferred "horizontally" from neighboring fields.[5]

In sum, contextualizing the lighting of matches and the borrowing of cash is difficult enough. Where do we begin capturing the contrast class of questions that presuppose a host of intra-field and inter-field scientific hypotheses, that is, inquiries that span within and across fields? Impossible as it may seem, students of science learn this quickly and relatively painlessly. How? The key is that, indispensable as they are, scientific questions are never introduced directly. The framing of an inquiry is always mediated by the specification of a *model*. Models are presented and discussed in detail in section 5.4. In the meantime, we can preliminarily characterize them as abstract, idealized systems that represent parts of the world on the basis of similarity. The claim that scientific explananda are contextualized by models is unlikely to raise many eyebrows. Still, this apparently unremarkable observation conceals dangerous pitfalls.

Intuitively, the framing of an explanandum must occur prior to the construction of a model. First, we decide what we are trying to explain. Next, we can construct a model to capture it and explain it. However, if the sharpening of the explanandum presupposes a model, we run into a classic "chicken and egg" problem. Posing the right explanandum in the appropriate theoretical context requires a model, and the model presupposes a phenomenon to be modeled. How does any of this get us off the ground?

An obvious solution is to posit that both endeavors must be addressed and developed simultaneously. What looks like two independent processes— framing and modeling—are really parts of a single endeavor. Framing an inquiry is part and parcel of the construction of a model. There is something true to this: models are goal-directed, they are conceived with an explanatory target in mind. At the same time, as just presented, this solution is a nonstarter, for the simple reason that models are costly. Questions—scientific and ordinary ones alike—have many possible readings. Each interpretation, on the view under present scrutiny, is represented by a different model, and pragmatic considerations dictate which one is most appropriate in the present context. This might work with relatively simple and intuitive scenarios, like the lighting of matches or Clyde lending cash to Alex. Yet, it would be preposterously inefficient to ask, say, a team of biologists to provide an array of models to frame various interpretations of the structure of a gene so that

---

[5] For a classic discussion of these "interfield theories," see Darden and Maull (1977).

the most appropriate one may be selected. Genetic models do not come easy! A viable solution must recognize the need for a cheaper, tentative scaffolding that provides an initial sketchy characterization of what we are trying to explain, informative enough to get the inquiry going, without making it overly expensive. How is this done without overburdening researchers with endless requests for alternative models?

This problem has a simple way out. Recall from Chapter 4 the concept of *frame*: a placeholder that stands in for a range of behaviors in need of explanation. Frames are the key to solving our conundrum. To specify an explanandum, one need not spell out an entire model. It is sufficient to provide the relevant "frame," that is, a shorthand description of the class of behaviors that the model will then set out to explain. This preliminary scaffolding, this coarse-grained characterization of the explanandum, is what kickstarts an exploration, suggesting how to move the inquiry further. Are these placeholders permanent, or are they progressively eliminated as science advances? This will be addressed in Chapter 10. For the time being, let me illustrate the main idea with an intuitive example. More realistic applications to actual scientific research are postponed until Chapter 6.

Why did Clyde lend Alex $300? Dretske's question is subject to interpretation. How does one determine whether the object of explanation is why Clyde lent Alex *$300*, why Clyde *lent* Alex $300, or why *Clyde* lent Alex $300? Garfinkel proposes an effective strategy to clarify the explanandum: specify a contrast class. Do we want to know why Clyde lent Alex $300 as opposed to $100 or $500, why Clyde lent Alex $300 as opposed to gifting the money, or why Clyde as opposed to Scrooge lent Alex $300? These coarse-grained characterizations are frames: placeholders that stand in for patterns of behavior, thereby sharpening our explananda.

Three brief comments: First, providing the appropriate frame, that is, specifying the contrast class, in and of itself, does not provide an answer to the question at hand. All it does is pinpoint *which* question we are trying to answer, which events are to be explained. Second, identifying the relevant contrast class clearly falls short of providing a complete model. A full-fledged model of why Clyde lent Alex $300 as opposed to $100 or $500 will specify why Alex needs the money, the relation between Alex and Clyde, how Clyde feels about Alex's financial situation, and much else. Yet, framing the question by specifying a contrast class is the first, preliminary step. It indicates the object of explanation, indirectly clarifying how the explanandum ought to be addressed and many other implicit features of the model. Third, the relation between placeholders and behaviors is best understood as one of representation. Patterns of behavior are concrete events in the world. In asserting that frames stand in for a range of behaviors, I am claiming that the placeholder provides a coarse-grained description—linguistic, pictorial, or otherwise—that characterizes what all these concrete events have in common, how they can be subsumed under a single causal explanation. In this representational sense, frames are placeholders which subsume actual or potential patterns of behavior in need of explanation.

With all of this in mind, we are finally in a position to present, more precisely, the first stage of our strategy. The construction of a black box begins by packaging and contextualizing the object of the explanation. Making the target fully explicit requires the development of a model, which is no trivial endeavor. Fortunately, in order to get an inquiry going, we do not need an exhaustive specification of the complete model. It is sufficient to provide a *frame*, that is, a preliminary coarse-grained description or general exploratory proto-model that stands in for the range of behavior in need of explanation. These placeholders accomplish what would otherwise appear to be an impossibly arduous task. They provide a preliminary scaffolding that allows us to present the explanandum quickly and efficiently.

§5.3. Step Two: The Difference-Making Stage

The first step toward the construction of a black box involves clarifying the structure of the explanandum. As we have seen in section 5.2, this is a complex endeavor, digging deep into the pragmatics of languages and theories. Fortunately, this aim can be achieved, in a fairly painless fashion, by providing a frame, a placeholder that stands in for the patterns to be covered. Once the inquiry is in focus, the next task is spelling out the explanation itself.

The topic of explanation has spawned a hefty literature that cannot be adequately reviewed here, let alone critically discussed. Given our present concerns, I shall focus on a prominent form of scientific explanation: *causal explanation*. How does one provide a causal explanation? The idea, simply put, is to explain an event by identifying the set of causes that produce it.[6]

---

[6] Here is David Lewis's (1986, p. 217) influential formulation: *"to explain an event is to provide some information about its causal history."* As the quote suggests, Lewis's contention goes further than mine. He maintains that *all* explanation is causal. I wish to remain agnostic on this controversial

This requires determining, first, which entities, structures, and processes make a difference to the target and, second, providing an apportionment of etiological responsibility. This is the essence of the next constitutive phase of the black-boxing strategy: the *difference-making stage*.

Let us begin with the nature of causal explanation. Recognizing that one way to explain an event is to provide information about its causal history is only the first step in a long hike. Any happening, even the most mundane, has a rich and dense history. This raises a question that Strevens (2008) calls the "problem of relevance." How do we determine which among myriad causes of an event should be included in its explanation?

One strategy to tackle the problem of relevance is to flatly deny that there is any principled distinction to be drawn between explanatory causes and non-explanatory ones. From this perspective—which Strevens calls the "minimalist account of causal explanation," or "minimalism," for short—there simply is no selection principle. Every causal influence is explanatorily relevant to the event brought about, no matter how distant in space or time.

A little reflection will convince most readers that minimalism, thus crudely stated, will not do. Every event is affected by a gargantuan net of causal influences literally tracing its roots to the dawn of time and extending to the corners of the universe. Is it not preposterous to allege that, say, the gravitational mass of Pluto and the Big Bang are part of the explanation of the event of me sipping coffee while typing these words?

Sophisticated forms of minimalism acknowledge this objection and attempt a response. Causal influence and, accordingly, explanatory relevance come in degrees. The current wintery weather and the aroma of freshly brewed coffee diffusing from the kitchen play the decisive role in prompting me to opt for this beverage. Pragmatic convenience dictates the omission of other causal factors, which are only marginally relevant, from abbreviated explanations. Nevertheless, the minimalist story goes, as the explanations get increasingly detailed and comprehensive, additional factors may be progressively included, encompassing Pluto and the Big Bang as limiting cases.

Prominent philosophers, such as Railton (1981) and Salmon (1984), have offered accounts of causal explanation along the lines of this sophisticated minimalism. Yet, Strevens maintains—correctly, in my opinion—that minimalism is too minimal. Specifically, it fails to deliver the intuitively correct

---

tenet. I assume the platitude that causal explanation is *one* form of explanation, without presupposing that all explanation is causal.

results whenever we have multiple potential causes of an event, but only one of them should be identified as *the* explanation of the event itself.[7] Any viable analysis of causal explanation must provide a principled distinction between explanatorily relevant factors and irrelevant ones.

How does one supplement minimalism with an appropriate selection principle? Strevens distinguishes two families of strategies. *One-factor approaches* develop a more nuanced concept of cause. From this standpoint, minimalism is correct that all causes explain. Still, not all influences are genuine causes. The cold weather and the pungent aroma are causes of my sipping coffee. Pluto and the Big Bang are not. A different path is followed by *two-factor approaches*, which agree with minimalism that all influences are causal, but stress that not all causes are explanatory. Pluto and the Big Bang might well be causes of me sipping coffee, or so the suggestion runs. Yet, an independent selection rule, a principle of causal relevance, will indicate that these causes play no role in the explanation in question.

Which theory of causal explanation best serves our purposes? Several options are available in the philosophical literature. The following is a non-comprehensive list. Regularity theories follow Hume in analyzing causal relations in terms of uniformities in nature. Statistical theories treat difference-making causes as factors that alter the probability of the explanandum. Counterfactual theories and manipulability theories view causes as conditions *sine qua non* for their effects: had the cause not occurred, the effect would not have occurred either. Finally, process theories and mechanistic theories view causes as physical connections, understood in terms of their capacity to transmit marks, exchange conserved quantities, or produce particular effects. Each strategy can be adapted to explanation in either the one-factor or two-factor guise. On the former reading, the theory specifies what counts as a cause of the explanandum event. On the latter interpretation, it suggests which causes are relevant to the explanation.

Strevens finds none of these accounts of relevance fully satisfactory. His own theory of explanation is founded on a different criterion for assessing the significance of causal factors. This is the "kairetic" recipe which, simply put, goes like this. Begin by taking a complete causal model of the production

---

[7] Strevens illustrates this with the story of Rasputin. Rasputin's assassins first attempted to poison him and failed. Next, they shot him and failed again. Finally, they successfully drowned him in a frozen river. Strevens considers several minimalist responses but argues that they all ultimately miss the mark. I concur and refer to him for further discussion. For structurally analogous scientific examples, see Nathan (2014).

of the explanandum, that is, the enormous description of all the causal influences leading up to the events to be explained. Next, make the description as abstract as possible without violating the following two conditions. First, it is important not to undermine the explanans' entailment of the explanandum, that is, one should not make the depiction so general that the explanandum-event no longer follows from the specification of causes and initial conditions. Second, the model must remain a causal model so that, for example, one cannot abstract away by replacing everything with a proposition to the effect that the explanandum occurred. If this process is applied correctly, what remains in the model after this gradual process of abstraction is completed are all and only the factors that actually made a difference to the occurrence of the event that we are trying to explain.

A simple example should help drive the point home. Consider a simple causal explanation: a piece of butter melts because it is heated. Note that a lot of properties will influence *how* the butter melts: its shape, weight, outside temperature, humidity, and many other factors. Yet, none of these features makes a difference to the explanandum, *that* the butter, in fact, melts. Therefore, they can be removed from the description without invalidating its status as a causal explanation. In contrast, the claim that the frying pan is heated above the melting point of butter cannot be abstracted away because it does make a difference to the occurrence of the effect under scrutiny.

So, which criterion or criteria of causal relevance should we employ to provide and assess causal explanations? Shall we opt for a regularity or statistical approach? A counterfactual or manipulability account? A process-mechanistic analysis? While each theory has its virtues and limitations, all of them will get the job done, one way or another. Fortunately, we are not forced to choose. I will not commit to any specific theory of causation or causal explanation. For the sake of illustration, I often borrow Strevens's kairetic account, that—setting aside the issue of whether and how it is possible, in practice, to perform the required operations of abstraction—provides a clear, simple, and compelling perspective. Nonetheless, I stress that nothing hinges on this particular choice. My analysis of black boxes can be paraphrased, *mutatis mutandis*, in terms of any theory of causal explanation, as long as it provides a recipe for identifying factors that make a difference to the outcome.

Next, what kind of entity produces the explanandum event? Depending on the specifics of the system under investigation, it may be a set of objects, events, processes, actions, or something else altogether. To avoid the tedious task of distinguishing and categorizing all these different phenomena, allow me to introduce a blanket term intended to cover all types and varieties of difference-makers. This is the concept of *causal mechanism*, or *mechanism*, for short. Mechanisms will be discussed in detail in Chapter 7. For now, note that they encompass a heterogeneous class of entities, which may occur at various scales, relative to the choice of explanandum.[8]

How does all of this relate to the construction of black boxes? Envision the situation from the perspective of Strevens's kairetic strategy. A difference-maker, on his view, is the result of a process of abstraction, which removes details from the description of the causal mechanisms that produce the explanandum in question. After all is said and done, what remains is a functional input-output description of the factors that play an ineliminable, crucial role in the occurrence of the effect. Had these factors been absent, the framed event would not have occurred. Thus described, difference-makers are placeholders. Specifically, they correspond to the type of placeholders introduced in Chapter 4 as—wait for it—"difference-makers"! They are abstract depictions that represent, in a coarse-grained fashion, the detailed mechanisms producing a specific behavior or pattern thereof.

It might be useful to further discuss the nature of these difference-makers. First and foremost, the amount of detail in these descriptions may vary. In some cases, the coarse-grained representation may specify a lot of detail concerning the structure of the underlying mechanism. Still, as long as it contains *some* abstraction, it constitutes a placeholder. At the other extreme, mechanisms can be referred to even when next to nothing is known about them. As Strevens (2008, §5.3) notes, even a virtually unidentified mechanism may be picked out as "the mechanism responsible for the behavior of a specific component of the system." This omission of detail—which Strevens aptly calls "black-boxing"—may occur for two different reasons and depending on the specific explanatory aims at play. Let us address them in turn.

---

[8] In some cases, causes and effects occur roughly at the same level, as when we say that the match lit because it was struck. But the causal mechanisms producing an event may also occur at finer or coarser levels. Consider accounts of the lighting of the match which appeal to chemical reactions triggered when phosphorus on the head reacts with potassium chlorate mix on the side of the matchbox. A precise characterization of levels of explanation is no trivial endeavor (Craver 2007). I appeal to an intuitive characterization of levels, on the assumption that causes must be "commensurate" to their effects.

On the one hand, the nature of the mechanism could be known. Nevertheless, one may choose to leave out details for the sake of pragmatic convenience. Suppose that you see fragments of ceramic scattered across my office floor. You ask what happened and I reply that I dropped my mug. I just provided you with a rudimental causal explanation. Pointing to the pieces of ceramic provides the frame. The dropping of the mug is the difference-maker. Could I say more? Of course. I could present both explanans and explanandum in greater detail by clarifying that it was my favorite mug, or that I dropped it accidentally on my way to class. With the help of some elementary physics, I could also tell you more about why and how the mug broke. Still, most interlocutors will be satisfied with my preliminary explanation. Additional information about the trajectory and the force of the mug hitting the ground, or my exact intentions, can be omitted for the sake of convenience, without affecting the adequacy of this coarse account.

Note that this is precisely the situation in Sober's *Drosophila* scenario. We can describe the relation between type-*a* and type-*b* flies in terms of fitness, although we could also provide a deeper explanation of the evolution of the population that dispenses with the concept of fitness altogether.

On the other hand, there are cases where the precise nature of the mechanism is unknown. It must therefore be left out as a matter of necessity. Consider sport-related discussions, common in bars across the globe. Your friend Jim is offering an unsolicited reconstruction of the last Superbowl, passionately explaining how the outcome was determined by coaching strategies. The winning coach's wisdom, the panegyric goes, was the difference-maker of the game. Jim can provide some evidence. He can pinpoint a few successful plays and offer a broad-brushed sketch of the strategy itself. Yet, the precise mechanisms that govern the amazingly complex relation between tactics and game results cannot be laid out precisely, by Jim or by anyone else. Here, the details are omitted not by choice, but by necessity. Does this mean that Jim's explanation is wanting? Not really. Assuming that Jim is, in fact, correct, about the difference-makers of the game, the explanandum has been causally explained. Sure, there are myriad other factors that have not been covered. Precisely which aspects of the coaching strategy are responsible for the outcome? How did the coach's decisions affect the game? Could the overall strategy be improved? These, however, are altogether different explananda. These questions are framed very differently, and thus require an altogether different causal explanation.

Scientific analogs of this latter situation can be found in the work of Darwin and Mendel, who were in no position to accurately describe the mechanisms underlying their explanations. More contemporary examples involve neurodegenerative diseases, such as Huntington or Alzheimer's, and certain types of cancer, whose molecular basis is only partially understood.

Time to take stock. All explanations need to be contextualized. Providing the appropriate setup is no easy task, especially in the case of scientific explanations that presuppose a host of inter-field and intra-field relations. Pragmatic convenience dictates that we employ a shortcut, developing some preliminary scaffolding that specifies the object of explanation in a fast and frugal fashion, while omitting lots of detail. This can be typically achieved by constructing a *frame*, a placeholder standing in for a range of behaviors in need of explanation. This is the first step of the black-boxing strategy, the "framing stage." The next step involves spelling out the causal explanation, a specification of the variables that make a difference to the target explanandum, thus contextualized. For this reason, I refer to this second phase as the "difference-making stage." The underlying causal mechanisms that produce the behavior in question are typically quite complex. The specifics of how mechanisms are implemented are often irrelevant. They can be omitted, because of either ignorance or convenience. The causal explanation provides abstract depictions of these mechanisms. I refer to such descriptions as *difference-makers*: placeholders standing in for the mechanisms that produce the behavior in question, as noted in Chapter 4.

We are now in a position to fully appreciate the importance of the framing stage in the causal-explanatory process. Consider the lighting of the match. Which factors should we enlist among the difference-makers, relative to this choice of explanandum? I will follow the kairetic recipe. Alternatively, feel free to pick your difference-making theory of choice. Begin with a complete specification of all the causal influences leading to the event in question. This web of influence will be enormous, including all sorts of minor disturbances. Next, make the description as abstract as possible, removing all the factors that do not affect the occurrence of the explanandum. Counterfactual reasoning dictates that, say, the gravitational pull of Jupiter can be omitted because, had Jupiter not been there, the match would still light. Similarly, we can take out the color of the matchbox, and many other irrelevant features. In contrast, the striking cannot be removed because, intuitively, had the match not been struck, it would not have lit. This shows that the striking is, indeed, a cause of the lighting.

Wait. Sure, had the match not been struck, it would not have lit. But had no oxygen been present in the atmosphere, the match would not have lit either.

And had the match been faulty, or made of plastic, it would also not have lit. This is Goodman's problem of counterfactuals.[9] Why do we treat striking as the cause of lighting, but not oxygen or the absence of manufacturing defects? The answer, once again, lies in the framing process, which will influence the selection of causes, the structure of the model, feeding back on the nature of the explanandum.

My central contention—the relativization of all causal statements to a frame of reference—is hardly novel. In his classic study, *The Cement of the Universe*, J. L. Mackie (1974, pp. 34–35) notes that causal statements "are commonly made in some context against a background which includes the assumption of some *causal field*."[10] Strictly speaking, then, what is caused is not an *event simpliciter*, but an *event relative to a causal field*. This has implications for our understanding of causal explanation. In assessing what caused event E relative to field F, some conditions that play a necessary role in the production of E can be dismissed as having no genuine causal role. These are the conditions that are part of F. But, if we evaluated the same statement relative to a different field, call it "G," then parts of F may now be said to cause E and, vice versa, some of the causes of E relative to F might become part of G, fading into the background.

Analogous considerations have been revamped by Strevens (2008). Echoing Mackie, Strevens treats the "given that" clauses which constitute the framework as part of a fixed portion of the explanation against which specific difference-makers are evaluated. More precisely, a state of affairs s is part of the framework F of explanandum E if all causal explanations of E must

---

[9] In *Fact, Fiction, and Forecast*, Goodman introduces the "problem of counterfactuals" as the task of defining "the circumstances under which a given counterfactual holds, while the opposing conditional with the contradictory consequent fails to hold" (1955, p. 4). Simple as this may appear, Goodman is quick to point out two formidable difficulties. A first issue, the "problem of law," challenges us to describe the nature of the systematic connection between the antecedent and the consequent of a subjunctive, given that such connection will typically not be a matter of logic but a natural, physical, or causal law. This endeavor is rendered thorny by the threat of circularity. Laws are counterfactual-supporting; accidents are not. Appealing to counterfactuals to identify laws and, simultaneously, employing laws to analyze counterfactuals generates a vicious circle or infinite regress. Goodman's second difficulty, more directly pertinent to our concerns, is known as the "problem of relevant conditions," or, alternatively, the "problem of co-tenability." The connection between antecedent and consequent in a counterfactual presupposes the occurrence of stable background conditions, often left implicit. "Had the match been struck, it would have lit" only holds provided that the match is well made and dry, oxygen is present, wind is absent, etc. Specifying *which* features must be taken in conjunction with the antecedent to infer the consequent is a long-standing philosophical problem. The connection between the context-dependency of explanation and subjunctives is noted by van Fraassen (1980).

[10] Mackie attributes the introduction of causal fields to John Anderson's "The Problem of Causation" (1938) and draws a connection to Russell's (1913) "causal environment."

contain *s*, because *s* makes a difference to the occurrence of *E*. Just as Mackie disqualifies any part of the causal field as a genuine cause, Strevens claims that *s* is not a difference-maker for *E*, as evidenced by the observation that *s* need not be mentioned explicitly in the causal explanation of *E*. In essence, in the context of a causal explanation of *E*, the status of *s* as "difference-maker" is suspended. It becomes a background condition.[11]

Technical as this may sound, the bottom line is a simple observation. All explanations, causal or otherwise, must be understood relative to a framework. And, although this is not invariably stated explicitly, the choice of the framework is always relative to the choice of explanandum. When I claim that the lighting of the match is caused by striking, I can dismiss the causal relevance of oxygen and other background conditions by including them in the framework. At the same time, when I reframe the question by asking why the match lit on a distant planet, the presence of oxygen becomes the difference-making factor and the striking event can be relegated to the background. This is why the framing stage is so important and why it is crucial to keep it distinct from the difference-making stage, where the causal explanation itself is spelled out.

With all of this in mind, we are ready to move on to the final step of black-boxing, the representation stage, where models re-enter the picture.

§5.4. Step Three: The Representation Stage

Are we done? Have we concluded our characterization of the black-boxing strategy? Not quite. After completing the first two steps, what we have is a causal explanation relativized to a frame. To obtain a black box, our framed causal explanation needs to be represented. We need to determine which features of our causal explanation should be explicitly described, and which can be omitted. This is done by embedding the contextualized causal story within a model. What is a model? How does it capture explanations? Section

---

[11] Two points are worth stressing. First, Strevens is more explicit than Mackie that frameworks can be introduced in discourse in a number of ways. It may be presupposed, implied, stated overtly, etc. Second, for Strevens, the introduction of a framework is optional. Many explanatory claims, he argues, are not relative to any framework and, therefore, they specify absolute as opposed to relative relations of difference-making. I respectfully disagree. I maintain that all explanations, no matter how simple or apparently unambiguous, presuppose a host of background assumptions. At the same time, my argument does not require this stronger claim. The main point, for present purposes, is the weaker observation that all *scientific* explanations presuppose a framework.

5.2 provided a general characterization of models as abstract and idealized systems representing features of the world. We now need to put some flesh on these bare bones and spell out a more developed account. This will provide the key to the final phase of black-boxing: the *representation stage*.

A good place to begin our discussion of models and representation is the influential work of Ron Giere. Giere's starting point is the observation that standard vehicles of scientific knowledge are not accurate descriptions of the world. What we find in textbooks and research articles across the sciences are highly abstract and idealized systems with no claim to reality beyond that conferred to them by the relevant community of researchers. Giere calls these constructs "theoretical models," or simply "models," for short.[12]

What is the relation between a scientific theory and the system it purports to represent? Traditionally, philosophers characterize the relation between theory and reality as a sort of correspondence. On this view, scientific theorizing provides "true" descriptions which accurately depict the corresponding portions of the world. Giere is adamant in rejecting this. For him, there is no direct connection between the world and its representations. The only relation is an indirect one, mediated by a theoretical model. Thus, what one finds in textbooks, what is generally called a "scientific theory," is best described as a cluster of abstract, idealized models, together with various hypotheses linking these constructs to real-world systems.[13]

Empirical hypotheses of the kind routinely found in science posit a relation of *similarity* between an abstract model and a physical portion of the world. Yet, since anything can be deemed similar to anything else, claims of similarity are vacuous without some more or less explicit specification of the relevant respects and degrees. To make the world-model connection more precise, Giere introduces the notion of "theoretical hypotheses," that is, statements of the form: "system S is similar to model M in the indicated

[12] Giere develops these views in his classic *Explaining Science: A Cognitive Approach* (1988). A similar perspective has been defended by Cartwright (1983), who maintains that the fundamental laws of modern physics, such as Schrödinger's equation, are not true. In her celebrated phrase, they "lie." Giere agrees with the content of Cartwright's proposal but opts for a different reformulation. For him, the general laws of physics cannot tell lies about the world because they are not statements about the world at all. They are, as Cartwright herself sometimes suggests, parts of the characterization of theoretical models, which may represent real systems. Here, I focus on Giere's presentation because his stance on models is more developed and explicit than Cartwright's. Be that as it may, the relevant issue, for our present purposes, is not much whether or not models "lie," but that they do not provide accurate representations.

[13] Contrary to the positivist conception of theories as interpreted axiomatic systems, discussed in Chapter 2, for Giere theories are not well-defined entities. No set of necessary and sufficient conditions determines which models and hypotheses belong to a theory.

respects and degrees." Unlike models, theoretical hypotheses are true or false, depending on whether or not the asserted relation actually holds. However, as Giere (1988, p. 81) is quick to point out, "that theoretical hypotheses can be true or false turns out to be of little consequence. To claim a hypothesis is true is to claim no more or less than that an indicated type and degree of similarity exists between a model and a real system. We can therefore forget about truth and focus on the details of the similarity. A 'theory of truth' is not a prerequisite for an adequate theory of science."

In sum, Giere characterizes models as abstract and idealized systems, vehicles for representing the world based on similarity. Giere was neither the first nor the last to recognize the importance of models in scientific practice. His main insight was replacing the old relation between model and reality in terms of correspondence with a relation of stipulation. Models are defined—created by their description. Consequently, whether a model sufficiently resembles the world is not a brute fact, but a collective decision of the members of the appropriate portion of the scientific community.

With this in mind, we can now consider, in greater detail, the issue of scientific representation. How do models represent their target systems? And how do we assess whether or not they succeed in doing so? Answering these questions will require us to discuss the anatomy of models.

Giere's account, originally developed with an eye to mathematics and physics, is difficult to apply to less formal branches of sciences, such as parts of biology, neuroscience, and psychology. For this reason, I will adopt a broader view of representation, which generalizes Giere's proposal while retaining its spirit. Weisberg (2013) characterizes models as *interpreted structures*. Unpacking this definition entails decomposing models into their two basic constituents: structures and interpretations.

Weisberg distinguishes three types of models—concrete, mathematical, and computational—based on the underlying structure. Concrete models are real physical entities that stand in some representational relation to some real or imagined system, the model's "intended target." Mathematical models employ formalized structures to represent states and relations between states. Finally, computational models represent causal properties of their targets by relating these causes to procedures. Setting these differences to the side, at the basic ontological level, all models are structures.

The second constitutive feature of models is their interpretation. Weisberg breaks down the interpretation of a model—what he calls the model's "construal"—into four parts: an assignment, the modeler's intended scope, and two kinds of fidelity criteria, "dynamical" and "representational."

The first two components of interpretation—*assignment* and *scope*—determine the relation between the model and real phenomena. As Weisberg (2013, p. 39) puts it, "Assignments are explicit specifications of how parts of real or imagined target systems are to be mapped onto parts of the model." This explicit coordination is important for two reasons. First, although parts of the model may map "naturally" onto specific portions of the world, this is often not the case. Second, and relatedly, it is not uncommon for assignments to be left implicit in discussions of models. The intended scope of the model, in contrast, specifies which aspects of potential target phenomena are represented by the structure and which are omitted.

A simple example may help clarify things. How do we use a concrete structure, beads on a string, to represent genes on a chromosome along the lines of what Ernst Mayr dubbed "beanbag genetics"(Figure 5.1)? The first step involves interpreting the structure. We have to specify, implicitly or explicitly, which aspect of the beads correspond to which properties of genes. Accordingly, an assignment must tell us whether each bead corresponds to a nucleotide or to a base-pair. It might further use the string to represent the backbone of phosphate and sugar residues keeping molecules together. Second, the scope tells us what is represented and what is not. Are the beads pattern-coded, that is, is there a signature mark for each of the four kinds of bases? Do the beads represent functional differences between sequences of nucleotides, such as coding sequences, operons, and promoters? Does the

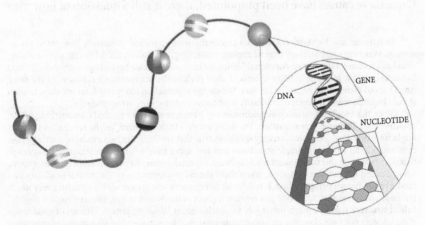

Figure 5.1. A simple model of genes: beads on a string.

model represent single strands of DNA or double helices? Are ribosomes included? Which properties convey information (number of beads, proximity) and which should not be taken literally (size, shape, texture)?

The third and fourth components of construal are its *fidelity criteria*. These determine how similar the model must be to the world, and in what respects, for the representation to be considered "adequate." Weisberg distinguishes two types of criteria. *Dynamical* fidelity criteria specify how close the output of the model, the values of dependent variables in the model and in the world, must be to the input of real-world phenomena. This is the "error tolerance" of the system, how far off predictions made on the basis of our beads on a string are from the behavior of real genes. The second family of fidelity criteria are *representational*.[14] Typically, these provide standards for evaluating how well the causal structure of the model maps onto the structure of the world, for the representation to be adequate. Do our large, colored, spherical beads provide a good enough approximation of real biochemical molecules? Can this simple structure capture the complexities of gene replication? What are the salient analogies and differences between the causal structure of the world and our representation of it?[15]

We are now in a position to state the final step of the black-boxing strategy. The representation stage determines which features of causal explanations should be portrayed and which can be idealized and abstracted away.

Was this not already done at the difference-making stage? Not quite. The goal of difference-making is, first, to identify the factors that make a difference to the occurrence of an event of choice and, second, to distinguish between genuine causes and conditions that may fade into the background. Once these causes have been pinpointed, there is still a question of how they

---

[14] As Strevens and DeMartino (personal communication) have independently brought to my attention, Weisberg's terminology may be slightly misleading, as dynamical fidelity can also be understood as fundamentally "representational." Since nothing of substance here hinges on the distinction between different kinds of "fidelity criteria," I shall stick to the original nomenclature. At the same time, I should stress that, regardless of how Weisberg's approach in interpreted, on my view, models should be understood in terms of similarity with target systems, not correspondence.

[15] Given this emphasis on interpretation, can any structure at all be a model? Can anything model anything else? To a first approximation, Weisberg answers in the positive. Yet, he recognizes that this might be misleading. While it is, strictly speaking, true that anything can be a model of anything else, concrete objects with very simple structures and few states have a "low representational capacity," that is, they are not able to represent many systems, especially ones of a very different type. Moreover, when dealing with concrete models, researchers should pay attention to the overall similarity of a model to the system being studied. It is difficult to represent one system with a structure very much unlike it. In this respect, while one can represent genes as beads on a string, a more complex double-helical structure will do a much better job. In conclusion, as Weisberg puts it, "The art of good modeling involves not only choosing an appropriate construal for a model, but also choosing structures with representational capacities appropriate to the target or the theoretical task at hand" (2013, p. 45).

should be represented. Precisely what goes in the model, and in what ways? Representation is achieved by constructing a model, an interpreted structure that represents parts of the world based on relations of similarity.

Following Weisberg, I distinguished three kinds of structures—physical, mathematical, and computational—and four aspects of interpretation. Assignment and scope map the structure onto the model. Dynamical and representational fidelity criteria determine whether the representation provides an adequate, informative characterization of the system under scrutiny.

This is all admittedly vague. Invoking the notion of similarity, in and of itself, does not resolve the issue of representation, which constitutes one of the thorniest and most resilient open problems in the philosophy of science.[16] Similarly, appealing to models does not, *ipso facto*, tell us how the explanandum should be represented and explained. Unfortunately, constructing and assessing models cannot be done from the proverbial armchair. It requires a painstaking combination of theoretical and empirical work. Still, the work of Giere and Weisberg provides an effective depiction of the representation stage. Metaphorically speaking, it involves constructing the framework that offers the biggest "bang for the buck" relative to the explanatory purposes at hand.[17]

Before moving on, a brief comment concerning the role of models in scientific inquiry. Models have been the object of much discussion in the philosophy of science.[18] Yet, traditionally, scholars have focused on a particular aspect of models, namely, their role as explanantia and explananda.[19] While models are a worthwhile object of investigation, and their contribution to the explanation of a phenomenon or range of phenomena is certainly significant, this is not the entire story. As we have seen throughout the chapter, models play a variety of other roles. For one, models frame the explanandum, making it more precise and perspicuous. Second, models provide the "toolbox" of explanantia. They specify the array of all the entities, laws, mechanisms, and

---

[16] For a comprehensive book-length discussion, see van Fraassen (2008).

[17] To the best of my knowledge, this colorful and effective metaphor was first used by President Eisenhower's Secretary of Defense, Charles E. Wilson, in 1954, to describe the "New Look" policy of depending on nuclear weapons, as opposed to a regular army, to keep the Soviet Union in check. More recently, it has been developed, in the context of scientific explanation, by Franklin-Hall (forthcoming), from which I borrow it.

[18] Suppes (1960) and van Fraassen (1980) are pioneers. More recent developments can be found in Cartwright (1983); Morgan and Morrison (1999); and Wimsatt (2007).

[19] This trend is slowly changing due, in no small part, to the pioneering work of Morgan and Morrison (1999) which emphasized the multifaceted role of models in science.

processes that can be employed in the construction of the explanans. Third, and finally, models provide the appropriate methodological standards. They tell us how the explanation at hand is supposed to operate and how to assess it. In essence, models are what allows researchers to address the challenge first posed by Goodman and subsequently developed by Mackie, Dretske, van Fraassen, Garfinkel, Strevens, and many others. Models effectively specify the background assumptions implicitly presupposed by an explanatory question and contextualize the inquiry. This framework does not precede the explanation. It is not implied by a theory. The framework is a constitutive part of the theory itself.

The radical philosophical implications of these commonplace observations will be explored, more systematically, in Chapter 7. For the time being, with all three steps in place, we have the conceptual resources to spell out the general definition of black boxes we have been seeking.

§5.5. What Is a Black Box? A Three-Part Recipe

This chapter has dissected the practice of black-boxing into three phases. The first step, the framing stage, involves sharpening the object of explanation. Specifying the explanandum in detail presupposes a full-fledged model, which is not easy to obtain, especially in actual scientific practice. Fortunately, to get the inquiry going, it is sufficient to employ a frame, that is, a coarse-grained placeholder that stands in for patterns of behaviors in need of explanation which, in principle, could be described at a finer scale. The second step, the difference-making stage, provides a causal explanation of the target by specifying which features of the explanans make a difference to the occurrence of the explanandum.[20] While there are various strategies for doing so, I borrowed Strevens's effective kairetic approach. Many of these difference-makers may be left unpacked. In some cases, the micro-structure is omitted because of mere convenience, to draw the boundaries of a field, or to insulate a concept from empirical refutation. Other times, the decision is dictated by ignorance, as these details are actually unknown. The third and

---

[20] As discussed at length, I do not have any novel account of causal explanation to offer and I wish to not take a stance on the question of whether all explanation is truly "causal." Still, it is worth stressing that the notion of difference-making adopted here is weak: it does not presuppose any physical interaction. Hence, my notion of black box can be applied to mathematical, statistical, and other forms of abstract explanation.

final step, the representation stage, determines how these difference-makers should be portrayed, that is, how much abstraction and idealization will produce the optimal explanatory "bang for the buck," given how the explanandum was framed. This is done by constructing a model, an interpreted structure that represents bits and pieces of the world based on a relation of similarity.

Our analysis began with a series of questions. What is a black box? How does it work? How is a black box constructed? What kind of entity is able to accomplish such a diverse set of tasks? We now finally have a skeleton of an answer. The three stages discussed in this chapter lead us to the general definition we were seeking. *A black box is a placeholder—frame or difference-maker—in a causal explanation represented in a model.*

A simple illustration is Putnam's geometrical explanation of the square-peg-round-hole scenario. The first step involves specifying the explanandum. The simple question "why does the square peg not go through the round hole?" introduces a frame. Contextualizing the object of the explanation makes the physical properties determining the rigidity of the board irrelevant. They can be assumed as part of the framework and, therefore, readily dismissed. The difference-makers are the geometrical properties of the system, mentioned in Putnam's higher-level explanation, which generalize to any rigid system satisfying the initial conditions. Still, reframe the explanandum—for instance, ask *what* makes the board and peg rigid—and the difference-makers will have to include some of these details.

Three comments. First, is this really it? Could we not have given this definition at the end of our historical excursus, sparing us two long chapters of dense philosophical analysis? We surely could have. But now we are in a position to fully appreciate the facets of the definition in question. At the most general level, a black box is a placeholder, a coarse-grained representation that stands in for entities whose nature and structure are either unknown or set to the side. As we saw in Chapter 4, there are two kinds of placeholders: frames and difference-makers. Correspondingly, there are two types of black boxes, depending on whether the representation in question spells out an explanans or an explanandum, that is, whether the description stands in for patterns of behaviors or for mechanisms. The causal explanation should be understood as a specification of factors that make a difference to the occurrence of the explanandum. The selection of these difference-makers is determined by the contextualization of the target-event itself, which is initially signaled by a frame and ultimately presupposes a scientific model, an abstract and idealized interpreted structure. The model also determines how these difference-makers should be represented—which aspects of the underlying mechanisms must be explicitly depicted, and which ones can fade into the background. When detail is omitted, this is done on the assumption that it does not affect the autonomy of higher-level explanations. No wonder it took us so long to cover all this ground. Phew!

Second, my characterization places both frames and difference-makers on their own, independent, and mutually exclusive spectrum. These placeholders range from general, coarse-grained depictions—"the mechanism responsible for genetic inheritance" or "the range of behaviors typically described as 'falling in love'"—to detailed descriptions of the underlying entities and activities. Hence, my general definition of a "black box" encompasses also less-opaque constructs, sometimes dubbed "gray boxes" or "semitransparent boxes," which shall be discussed, at greater length, in Chapter 7. What all these placeholders have in common, from my perspective, is that they result from a process of abstraction and idealization.

Third, as anticipated in section 5.1, these stages are necessary and jointly sufficient for the construction of a black box. Not all valuable scientific work requires all three phases. But all black boxes do.[21] Yet, these steps should not be viewed as chronologically ordered. Sure, there is a natural progression which begins by framing the explanandum, then provides the explanation, which is finally represented in a model. Elegant as it is, this conception of "well-ordered science" is just a regulative ideal. Actual practice is messy. Research teams often begin by hypothesizing a causal explanation, which is then found wanting, leading to the revision of the explanandum. Or a revised representation of difference-makers might lead to a revisitation of the explanandum, which, in turn, changes the nature of the causal explanation. In short, the implementation of these steps may involve permutations, repetitions, and feedback loops. All of this is captured by Firestein's (2012, p. 19) gloss: "'Let's get the data and then we can figure out the hypothesis,' I have said to many a student worrying too much about how to plan an

---

[21] Some readers might accept that all three steps are required to construct a black box, understood as a difference-maker. But why does the construction of a black box *qua* frame require difference-making and representation? Is the framing process not sufficient? My reason for answering in the negative is simple. While a frame, per se, is a preliminary exploration, a black box is always relativized to an entire model. And this requires the processes of difference-making and representation, in addition to framing.

experiment"—incidentally, an example of data mining which, like black-boxing, is praised by some scholars while frowned upon by others.[22] Still, all three phases are required to characterize how black boxes are constructed, used, and assessed in science. Speaking of which. . . .

We will get back to these questions and related issues of explanatory reduction in Chapter 10, where we shall also address the question of whether scientific explanation will invariably incorporate some black boxes, or if all these constructs can be progressively opened and eliminated, at least in principle. Meanwhile, so much for square pegs, matches, cash lenders, bank robbers, and other toy examples. Does our general definition capture the use of black boxes in real historical cases? What implications does this have for philosophy at large? It is time to put the recipe to work. Chapter 6 applies our three-step recipe to Darwin, Mendel, Skinner, and Friedman. Next, we shall offer some philosophical payoff of the strategy at hand.

[22] Broadly speaking, data mining can be characterized as the practice of examining large databases in order to generate new information. Data mining is controversial because it tends to seek out large-scale correlations and set aside any causal-mechanistic underpinning.

6

History of Science, Black-Boxing Style

The history of science, after all, does not just consist of facts and
conclusions drawn from facts. It also contains ideas, interpretations of
facts, problems created by conflicting interpretations, mistakes and so
on. [. . .] This being the case, the history of science will be as complex,
chaotic, full of mistakes, and entertaining as the ideas it contains and
these ideas in turn will be as complex, chaotic, full of mistakes, and entertaining as the minds of those who invented them.

—Paul Feyerabend, *Against Method*, p. 3

§6.1. Introduction

The previous chapter provided an analysis of the black-boxing strategy by
breaking it down to three constitutive steps. First, the framing stage sharpens
the explanandum by contextualizing the object of explanation. This gets
the inquiry going by constructing a frame, a placeholder that stands in for
patterns of behavior to be covered. The second step, the difference-making
stage, provides a causal explanation by identifying the factors that make a
difference to the occurrence of the explanandum: its difference-makers.
The nature and structure of the mechanisms producing the phenomenon
under scrutiny are often quite complex. But not all details are required or
even useful for the explanation which, when appropriately framed, may well
be perfectly autonomous. Which aspects must be explicitly represented?
Are there features that may be idealized or abstracted away? What are the
standards for the explanation? Which assumptions can be presupposed, and
which need to be introduced and defended explicitly? These questions are
addressed in the third, and final, representation stage, which aims at producing a model of the explanandum. The significance of this step is often
neglected in discussions of causation and causal explanation. It is common
to talk as if causal attributions are absolute or context-independent. This is
not the case. As our discussion made clear, explanations—and, especially,

Black Boxes. Marco J. Nathan, Oxford University Press. © Oxford University Press 2021.
DOI: 10.1093/oso/9780190095482.003.0006

scientific explanations—are always relativized to a frame of reference. Consequently, causal claims are mediated by models that determine how the explanandum should be represented, explained, and ultimately assessed.

These three steps jointly suggest and support a general definition. A black box is a placeholder in a causal explanation represented in a model.

Chapter 4 distinguished two types of placeholders: frames and difference-makers. Accordingly, we have two types of black boxes, depending on whether the placeholder is an explanandum standing in for patterns of behaviors or an explanans representing the mechanisms producing the behaviors in question. The causal explanation spells out the difference-makers, which are then represented in a model, an interpreted structure that stands in a relation of similarity with bits and pieces of the real world. The nature and structure of these black boxes—which may also be "gray" or "semi-transparent"—vary substantially, depending on the amount of detail included. But they all have in common the omission of content, idealized or abstracted away for pragmatic convenience or necessity, on the assumption that this does not affect the autonomy of the higher-level explanation.

Time to put this strategy to work and address the following questions. Is our analysis adequate? Is it accurate? Does it capture the essence of black-boxing? What are its limitations? At the end of the previous chapter, the definition was illustrated by a toy example: Putnam's square-peg-round-hole system. Now, let us look at some real scientific applications.

At a general level, the second half of this book focuses on two broad issues. First, does our three-step recipe for constructing black boxes fit in with the vagaries of the case histories first introduced in Chapter 3? Now, given that our three-step recipe for black-boxing was motivated and set up by discussions of evolutionary theory, classical Mendelian genetics, behavioristic psychology, and neoclassical economics, it may not come as a shock to see, in the present chapter, that the analysis in question successfully captures the same episodes that inspired it. Nevertheless, our discussion of black boxes does not merely strive to vindicate the initial choice of examples. It puts them in a different light. Second, what theoretical implications does all of this have? Presenting and assessing the philosophical payoffs of the black-boxing strategy will be the aim of the chapters to follow.

Before moving on, I should stress, once again, that my goal is not to provide a comprehensive narrative of these momentous events in the history of science. This substantial endeavor lies beyond the scope of this work. More modestly, my objective is, first, to provide an accurate rational reconstruction of the role and significance of black boxes across scientific practice and, second, to discuss its philosophical applications and implications.

§6.2. Darwin's Black Boxes

Darwin's theory of evolution by natural selection provides a powerful apparatus for explaining the distribution of organisms and traits across the globe. *On the Origin of Species*, can be viewed as one long argument organized into three main parts. The first four chapters of his masterpiece present the fundamental tenets of his evolutionary theory. These are the ingredients skillfully blended by Darwin into the main course: the principle of natural selection. Chapters 5–8 provide evidence for natural selection, anticipate objections, and respond to them. Chapters 9–13 put the account to work, revealing what can be accomplished by applying these insights to problems in biology. The final chapter summarizes and wraps up his discussion.

Let us focus on the third section of the *Origin*, where the breadth and depth of evolutionary explanations truly come to life. Specifically, consider Chapters 11 and 12, which address issues of biogeography. There, Darwin begins by considering three striking facts regarding distribution patterns. First, neither similarities nor dissimilarities among organisms can be fully accounted for by environmental features alone. The reason is evident. Species inhabiting distant lands differ vastly and, yet, continents across the globe display similar climatic conditions. A second observation concerns how geographic barriers, as well as other obstacles to free migration, deeply affect variation among organisms and ecosystems. A clear illustration is the difference in marine flora and fauna inhabiting eastern and western shores of Central America, attributable to the narrow albeit (for many species) impassable isthmus of Panama. Third, there are unquestionable analogies between species inhabiting similar environments. Organisms from the same continent or ocean share features that set them apart from species in other areas of the planet, as shown by the uniqueness of Australasian marsupials. These three facts, Darwin remarks, call for an explanation.

Given the remarkable scope of the explanandum—worldwide patterns of geographical distribution—one could legitimately expect Darwin's analysis to be rich in detail and structure. This expectation is misplaced. Readers of the *Origin* are presented with a surprisingly minimal account:

We see in these facts some deep organic bond, prevailing throughout space and time, over the same areas of land and water, and independent of their physical conditions. The naturalist must feel little curiosity, who is not led to inquire what this bond is. This bond, on my theory, is simply inheritance, that cause which alone, as far as we positively know, produces organisms quite like, or, as we see in the cases of varieties, nearly like each other. The dissimilarity of the inhabitants of different regions may be attributed to modification through natural selection, and in a quite subordinate degree to the direct influence of different physical conditions. (1859 [2008], pp. 257–258)

Thus, in a few short statements, Darwin provides the key to one of the great mysteries of science: the question of biogeography. How is this possible?

Some readers might feel inclined to downplay the brevity of Darwin's analysis by appealing to its obviousness. From a historical perspective, this would be a mistake. Darwin's view, deeply rooted in contemporary biological thinking, might sound like a truism to a modern audience. Yet, one should not ignore the originality of the proposal and the stark opposition confronting evolutionary theory since its inception, and which continues inexorably to this day, at least in some circles. Darwin's insight challenged some of the most basic and entrenched, but still widely debated, credence of its age. These include the belief that organic diversity is the result of divine creation, that the Earth is much younger than it is, that natural phenomena must be explained via teleological concepts like design, and an anthropocentrism placing humans at the pinnacle of the *scala naturae*. In addition to his contribution to discrediting these hallowed ideas, Darwin also bolstered a number of key concepts, such as the replacement of essentialism with population thinking, the principle of natural selection, geographic speciation, and the understanding of evolution as a process. Some will surely quibble with Mayr's somewhat hyperbolic conclusion that "no other philosopher or scientist has had as great an impact on the thinking of modern man as Darwin" (1988, p. 194). Yet, it is hard to overstate how evolution by natural selection constitutes a traumatic rupture with the worldview of orthodox Christians, natural theologians, laypeople, as well as many philosophers and scientists born and raised in the nineteenth-century intellectual milieu.

These considerations bring us back to our original question. Given this complex thicket of conceptual innovation and intellectual battles, how is it possible for Darwin's evolutionary analysis to be so beautifully simple? How can such a succinct explanation do so much theoretical heavy lifting?

The answer, in a nutshell, is that Darwin achieves such an elegant explanation through his effective construction of black boxes. Allow me to elaborate by decomposing Darwin's endeavor into our three key steps.

To begin, Darwin's target is not biogeography *simpliciter*. He is after a highly hedged and contextualized set of patterns. To wit, myriad distributions of organisms can be observed on the planet. Not all of them count as an appropriate explanandum for evolution by natural selection. Darwin's analysis can account for differences between, say, cave-dwelling animals on different continents and anatomical similarities between Australasian marsupials. Yet, it does not apply to distributions that are the effect of luck, such as the outcome of natural catastrophes. It does not cover the production of traits that do not differ with respect to fitness, or the origins of life.[1] In short, Darwin's explananda are far from absolute, self-standing events. They are specific problems, relativized to contexts and frameworks.

What is Darwin's explanatory model? Simply put, it encompasses his "theory" of evolution by natural selection. This corresponds to the apparatus patiently spelled out in the initial chapters of the *Origin*, that we succinctly reviewed in section 3.2 of Chapter 3. The existence of a process of natural selection can be directly inferred from the struggle for existence described by Malthus, together with the ubiquitous presence of variation in nature, some of which affects fitness and can be inherited across generations. Such textbook reconstructions should not be taken too literally. Breaking the theory of evolution down to a handful of axiomatic principles fails to do justice to its groundbreaking contribution. What Darwin describes as "one long argument" also involves a number of methodological directives, such as the adequacy conditions that he borrows from Herschel, Lyell, and other precursors. It includes painstaking observations—some novel, others well known—like the paucity of the geological record or comparisons between organisms and environments across the planet. Finally, it encompasses more or less implicit assumptions concerning what he deemed obvious, preposterous, or worthy of exploration—Grene and Depew (2004) call these "epistemic presuppositions"—together with objections and his responses. This

---

[1] To be clear, I am *not* arguing that contemporary evolutionary theory has nothing to say about chance or the origins of life. It does. My point is that Darwin was is no position to answer these questions. In this sense, the title of his masterpiece is misleading. He offers lots of insights on the evolution of species, but not much about their "origin."

hodgepodge of assumptions, observations, anecdotes, side remarks, and hypotheses provides the backdrop against which Darwin's research must be understood, processed, and assessed. In a nutshell, they are his model. Parts of this evolutionary theory are stated explicitly. Still, if Darwin were to spell out in full the host of intra-field and inter-field connections that lurk in the background, his inquiry would never take off the ground. Many, indeed most, premises are implicit. Thus, when the English naturalist states that he shall consider "the distribution of organic beings over the globe," this coarse-grained description is a frame: a placeholder standing in for specific, highly hedged distribution patterns in need of explanation. These frames are what allow his inquiry to be presented in such a simple, succinct, and elegant fashion.

The framing of the explanandum concludes the first stage of black-boxing. Next is the difference-making phase, where the causal explanation is laid out. Which variables make a difference to the production of the patterns of geographical distribution presented in the *Origin*? My rational reconstruction borrows from Strevens's kairetic recipe.[2] As noted in section 5.3 of Chapter 5, similar results can be obtained by following various alternative strategies. Begin with a detailed description of a specific distribution subsumed under the frame discussed earlier. For instance, consider some differences in the marine flora and fauna along the eastern and western shorelines of the isthmus of Panama. Next, make the depiction as abstract as possible without undermining the entailment of the explanandum, that is, ensuring that the distribution in question still follows from the relevant initial conditions and law-like generalizations. What are the decisive factors in producing the patterns under scrutiny? The short answer is descent with modification.

Obvious as all of this may sound to contemporary readers, it was quite controversial in the mid-nineteenth century. Prominent intellectuals still believed that the only plausible answers would appeal to independent creation or intelligent design. Other influential minds considered this question impossible to answer—Lyell's "mystery of mysteries." Darwin had a different solution in mind, as we saw in the passage quoted earlier: "The dissimilarity of the inhabitants of different regions may be attributed to

---

[2] I refer to this as a "rational reconstruction" because, to the best of my knowledge, Darwin himself never explicitly identifies these three steps. Still, reformulating the explanation in contemporary terms allows us to identify and characterize its distinctive aspects.

modification through natural selection, and in a quite subordinate degree to the direct influence of different physical conditions." *Ecce* Darwin's causal explanation.

The abstract nature of the difference-makers renders the analysis perfectly general. Hardly specific to the Panama coastlines, it applies, *mutatis mutandis*, to organisms and environments across the globe. Why are native inhabitants of the Galapagos similar to the fauna of South America and different from species living on islands and mainland elsewhere? Darwin answers in terms of a history of migration from the continent with different adaptation to local environments explained by gradual inheritable differences in fitness. Why are the forelimbs of whales and apes functionally distinct but structurally similar? Again, descent with modification is key.

All this is now commonplace. Still, Darwin had numerous hurdles to overcome, in order to convince skeptical audiences. How could plants and other less-than-mobile species cross large bodies of water to colonize distant lands? How does one explain discontinuous distributions of organisms, such as trees flourishing on mountaintops, but absent in between? Darwin's willingness to acknowledge these and similar problems displays his intellectual honesty. His ingenious solutions reveal his talent for natural history.

Is that it? Is this all there is to Darwin's contributions to evolutionary theory? Not quite. The reason is that Darwin's causal explanation, as just presented, is too coarse-grained to be persuasive. For one thing, it is not perspicuous enough. Let us grant him that natural selection is the key difference-maker. What is descent with modification? How does it account for geographical distribution? What are the main forces and mechanisms at play? Unless we provide some answer to these questions, Darwin's proposal cannot be distinguished from other proto-evolutionary accounts, such as the theories offered by Lamarck and by his own grandfather, Erasmus Darwin. In short, appealing to an unqualified notion of "natural selection" was not going to cut it. And Darwin was well aware of this.

Clearly, not all aspects could be described in detail. As discussed, there was much that Darwin did not know. He was not in a position to pinpoint the mechanisms of inheritance, variation, fitness, and many other core features of his own theory. Even if he could, that would have made the explanation cumbersome and overly specific. This raises important questions. Which aspects needed to be made explicit? Which could be left tacit? How does one denote mechanisms whose nature and structure are unknown or too complex for the explanation at hand? Solving these problems is the task of the representation phase of the black-boxing strategy.

The key to Darwin's representation is his preliminary characterization of the theory. In the initial chapters of *Origin*, the English naturalist breaks down evolution by natural selection to its basic constituents: competition, variation, fitness, and heritability. These are all placeholders in a causal explanation represented in a model. These are Darwin's black boxes.[3]

To illustrate, recall from section 4.2 of Chapter 4 that fitness plays a dual role in evolution. On the one hand, it is a frame that captures distributions of organisms and traits. On the other hand, it is a difference-maker that stands in for mechanisms producing the distribution in question. Analogous considerations apply to competition, variation, heritability, and other key constituents of evolutionary theory. All of them figure prominently as placeholders, frames, and difference-makers, in Darwin's evolutionary explanations.

In conclusion, are Darwin's explanations successful? Generations of biologists agree on a positive answer and it is not hard to see why. Darwin framed the right questions about the evolution of species, identified the correct difference-makers, and appropriately represented their salient aspects in his model. Sure, there is much that Darwin did not know or even got plainly wrong. At the time *Origin* was published, Darwin was ignorant about the mechanisms of variation and inheritance, and his subsequent speculative theory strikingly misses the mark. Nevertheless, ignorance and mistakes do not affect his main accomplishments. How can this be so?

We now have a sketch of an answer. The model does the work expected from a scientific theory: framing, explaining, and representing phenomena. Two kinds of placeholders allow Darwin to provide an explanation that is simple, elegant, and concise. Frames pick out patterns of behavior needing to be covered. Difference-makers stand in for mechanisms that produce these patterns, regardless of whether these processes are known, unknown, or partly identified. These placeholders are Darwin's black boxes.

---

[3] In choosing my words, I am perfectly aware that the expression "Darwin's black box" is sometimes used as a scornful dismissal or undermining of Darwinian explanations (Behe 1996). My perspective is diametrically opposite. As Mayr promptly recognized, Darwin's black boxes are what *enhance* his explanations, in the face of what the great naturalist could not know.

§6.3. Mendel's Black Boxes

At the turn of the twentieth century, biologists noted some interesting patterns emerging in experiments involving plant and animal breeding. The phenomena in question can be observed when two varieties of organisms displaying some observable phenotypic characteristics are crossed to produce viable offspring. For instance, consider a red-eyed fruit fly mating with a white-eyed fly, or a yellow-seed pea plant crossed with a green-seed plant. In the first generation, one of the two parental characteristics disappears, with all the offspring consistently showing the other one. Specifically, in the examples at hand, all fruit flies will be red-eyed, and all pea plants will be yellow-seeded. However, if these first-generation offspring are then crossed with one another, the second generation will display both characters present in the two original varieties. Moreover, these traits will reappear in a specific predictable ratio, namely, 3:1. In our examples, three-quarters of the second-generation flies will have red eyes and the remaining quarter will have white eyes. Similarly, three-quarters of the second-generation pea plants will be yellow-seeded, and one quarter will be green-seeded.

There is a compelling explanation of these and similar phenotypic patterns. Each organism contains two "factors" that determine which observable trait will be displayed. One factor comes from each parent and, if an organism inherits two different factors, one is always expressed preferentially over the other. These characters remain within it for life and are passed on virtually unchanged to the next generation. Today, grade school kids are familiar with these factors, which are now called "genes," a term introduced by the Danish geneticist Wilhelm Johannsen in the early 1900s.

The preceding patterns were first observed and explained, along the lines just described, by Mendel in a paper published in 1866. Hence, we can refer to them as "Mendelian ratios."[4] Contrary to common wisdom, these results were published in a respectable journal and were fairly well known among scientists working on related problems. Yet, these observations assumed a much broader significance at the dawn of the new century. Early geneticists such as Bateson and Johannsen soon realized that Mendel had provided the key to unlock Darwin's mystery: the basic principles of heredity.

---

[4] I borrow this expression, and much else, from Griffiths and Stotz (2013). Excellent discussions of classical genetics can also be found in Allen (1975, 1978); Darden (1991); Morange (1998); Waters (2004, 2006); Weber (2005); Falk (2009); and Dupré (2012).

As we saw in section 3.3 of Chapter 3, Mendel and his twentieth-century followers—who are now called "classical geneticists"—had a better grasp of biological inheritance compared to most naturalists of the time. Still, this deeper understanding boils down to more accurate laws and law-like generalizations, which constrain the structure and behavior of phenotypic patterns. From a mechanistic standpoint, the Bohemian geneticist black-boxed the underlying entities and processes responsible for these patterns no less than his illustrious English colleague. This leads us back to our original question. How did classical genetics advance the study of inheritance while lacking a substantial grasp of the underlying cytological workings? The answer emerges as soon as we break down Mendel's use of black boxes to our three steps.

To begin, Mendelian ratios are not universal. For one, they only apply to inherited traits and not, for instance, to acquired, learned, or otherwise environmentally induced characters, like sunburns or language learning. More importantly, not all innate phenotypic features display the patterns in question. Height and size do not follow the same rules. How fortunate of Mendel to randomly stumble upon the right kind of traits in the right kind of organisms. . . . Actually, the success of Mendel's experiments was hardly the product of luck at all. Due to his profound knowledge of botany and related areas of biology, the Bohemian monk was perfectly aware of the importance of selecting the appropriate plants for his inbreeding experiments.[5]

In his seminal 1866 article reporting his original findings, Mendel was explicit that, to avoid the possibility of jeopardizing the chances of success from the very beginning, the experimental plants must possess the following features. First, they must display constant differing traits. Second, their hybrids must be protected from the influence of all foreign pollen during the flowering period, or easily lend themselves to such protection. Third, there should be no marked reduction in the fertility of the hybrids and their offspring in successive generations. And finding the appropriate plants was only the first step. Next, Mendel had to choose the right variety and, finally, select suitable traits. This painstaking research is described well by Mayr:

---

[5] Some scholars have raised doubts about Mendel's results being "too good to be true." Assessing claims of Mendel's data being "fabricated" transcends my historical competence. Fortunately, from our present standpoint, settling this issue is not necessary to establish my main philosophical point.

Mendel procured 34 more or less distinct varieties of peas from several seed dealers and subjected them to two years of testing. Of these varieties, 22 remained constant when self-fertilized, and these he planted annually throughout the entire experimental period. In these 22 varieties, seven pairs of contrasting traits were chosen for experimental testing. Two plants differing in a given pair of traits were hybridized, and the behavior of the trait was followed in the ensuing generations. The 22 varieties differed from each other by far more than the seven selected traits, but Mendel found the other traits unsuitable because either they produced continuous or quantitative variation not suitable for the study of the clear-cut segregation that he was interested in, or else they did not segregate independently. (Mayr 1982, p. 714)

This shows that, when Mendel talks about "experiments in plant hybridization," the expression is what I call a "frame." It is a coarse-grained placeholder that stands in for a particular range of behaviors in need of coverage. The object of explanation is not inheritance *simpliciter*. The range of explananda is carefully selected and embedded in a specific framework.

This point can be reinforced by looking at how Mendel's own work was subsequently developed in the twentieth century. Mendelian ratios are typically presented and explained by appealing to two of Mendel's laws: the *law of segregation* and the *law of independent assortment*.[6] This reconstruction, however, is not quite accurate, from a historical perspective. This is because, right after the "rediscovery" of Mendel's work, the two laws were not considered conceptually distinct. Initially, the focus was on segregation. Independent assortment was viewed as a simple extension of segregation to two-character cases or "dihybrid crosses." This view was eventually abandoned when exceptions to independent assortment began to appear.

In 1905, Bateson, Saunders, and Punnett were able to confirm Mendel's law of segregation for various characters in sweet peas. At the same time, when they tried to perform two-character crossings with purple vs. red flowers and long vs. dry pollen, they encountered deviations from the 9:3:3:1 ratios. Simply put, the problem was that the traits "purple" and "long" displayed a tendency to be inherited together, and so did "red" and "round."

---

[6] As noted in section 3.3 of Chapter 3, paraphrased in contemporary terms, the *law of segregation* states that each parent passes on only one allele to each offspring. The *law of independent assortment* says that which allele an organism gets from a parent in one locus has no effect on which allele it gets in another locus. Mendel's third law, the *law of dominance*, tells us that, for every pair of alleles, one allele is "dominant," the other "recessive." Organisms who have a copy of each will look just like organisms that have two copies of the dominant allele.

The explanation of this partial coupling was eventually provided by T. H. Morgan, who treated it as a mechanical result of the location of the factors of the chromosomes—an idea that fueled the genetic technique now known as "linkage mapping."

Setting details aside, the point to be appreciated is the significance of the choice of explananda. *What* we are trying to explain is just as crucial as *how*. All these explanations are relativized to a context. Mendel's background model is constituted by his unique blend of the mathematics, physics, and botany of his time. The model sketched by Bateson, de Vries, Johannsen, Morgan, and other twentieth-century Mendelians was eventually molded into the "theory" of classical genetics. Such choices of explananda can only be understood against these theoretical and experimental backdrops, much of which is left implicit. This is why the framing stage is so important.

We now have the resources to move on to the other two stages of black-boxing: difference-making and representation. Beginning with the former, Mendel does not attempt a mechanistic explanation, in the contemporary sense of the term. As noted in section 3.3 of Chapter 3, he does not provide a detailed description of the process of heredity. There are two reasons for this. On the one hand, he was not in a position to describe the phenomena in question. Furthermore, he had an altogether different goal in mind. Influenced by his training in mathematics and physics, Mendel sought a characterization of the laws, not the mechanisms of variation. From this perspective, we can view Mendel's laws as his causal explanation of heredity, framed along the lines discussed earlier. The nature of the main difference-makers should now become clear. What makes a difference in heredity are those entities which Mendel called "factors" and Johannsen dubbed "genes." This is the backbone of the explanation provided by Mendel and subsequent classical genetics. Genes, thus construed, are black boxes of the second type. They are difference-makers.

This leads us to the third stage. What is there to be represented? The explanation of heredity in terms of Mendel's laws has become entrenched in contemporary thinking. Thus, it is easy to overlook how innovative and controversial it was and how much it differed from competing accounts. A brief overview of the main players on the field will reveal the significance of the framing and representation of these causal explanations.

One strategy to explain inheritance was to postulate some physical mechanism underlying and governing the process. We have already encountered a proposal along these lines. Darwin's "provisional hypothesis of pangenesis," outlined in section 3.2 of Chapter 3, followed precisely this route. Gemmules were introduced as the particles responsible for the transmission of heritable qualities. A more sophisticated version of this approach was offered, a few years later, by the German embryologist August Weismann, who approached inheritance from the perspective of developmental physiology. Specifically, he postulated an elaborate hierarchy of hereditary units that control ontogeny, ranging—in ascending order of size—from "biophores," to "determinants," to "ids." This was based on the insight that all genetic material must be contained in the nucleus. Incidentally, this rejection of "soft" inheritance of acquired characters is what conclusively discredited Darwin's pangenesis.

An altogether different approach came from biometrics. Biometricians, such as Darwin's half-cousin Francis Galton, believed that the aim of studying heredity was to discover the statistical relation between features of an organism and features of its offspring. Along these lines, Galton's "law of ancestral heredity" stated that one-half of the characters of an individual is explained by the characters of its parents. One-quarter is explained by the grandparents, one-eighth by the great-grandparents, and so forth. Karl Pearson, another eminent biometrician, went further, explicitly stating that Galton's law was not a biological "law" at all. Ancestral heredity, Pearson maintained, is a statistical measure that could be used to predict any variable from a set of variables with which it is correlated. Galton's weights are thus only estimates. Actual values must be determined empirically. Influenced by Mach's instrumentalism, Pearson viewed the aim of science as the systematization of observation. Any attempt to go beyond what is directly observable diverts us away from science, delving into the realm of metaphysics. From this standpoint, Pearson viewed Darwin, Weismann, and Mendel as guilty of the same mistake: introducing unobservable mechanisms of heredity as objects of research.

Classical genetics, early in the twentieth century, followed a third route, distancing itself from both previous proposals. Mendelians rejected Darwin and Weismann's focus on material particles as overly speculative. At the time, there was no direct, tangible evidence for such entities. In turn, they dismissed Galton's and Pearson's focus on observable patterns as simply wrongheaded. In Johannsen's terms, what organisms inherit are their "genotypes," not their "phenotypes." This being so, the relation between the phenotypes of parent and offspring is not the real phenomenon of heredity. It merely provides evidence that can be used to investigate the relation between the genotypes of the two organisms. Johannsen clarified this by introducing the term "gene," referring to the entities which constitute the organism's genotype. Each gene comes in a variety of alternative forms ("alleles") and an organism contains a number of places ("loci"), one for each gene. Thus, a genotype is the combination of all the pairs of alleles at that locus.

So, what does all of this mean? The heart of Mendelian genetics—the gene—has a distinctive status. It is not observable. But it is more than a mere unobservable posit to explain data. Mendel's factors are a tool for predicting and explaining Mendelian ratios in breeding patterns. They may stand in for these patterns of behavior or, alternatively, for the underlying mechanisms that make a difference to these variants. It was only natural for many geneticists to hope that the gene, thus construed, would eventually be shown to exist. Yet, as T. H. Morgan noted in his 1934 Nobel lecture, the centrality of genes in early genetics did not depend, in any significant way, on their status as physical particles. This, in short, is the theoretical role of genes, their representation in the model of classical genetics.[7]

A comprehensive historical reconstruction lies beyond the scope of this work. From our perspective, the important point is that genes are essentially placeholders in a causal explanation represented in a model. Genes have a multifaceted nature. They may stand in for various mechanisms or patterns of behavior. They can be frames or difference-makers. Accordingly, these details may be integrated in different ways or appropriately omitted. This is why they play so many different roles. Genes are Mendel's black boxes.

§6.4. Skinner's Black Boxes

Behaviorism is often presented as an uncompromising attempt to transform psychology from the pre-scientific study of consciousness, into a systematic empirical inquiry meant to explain, predict, and control behavior.

---

[7] The historian Raphael Falk (2009) sums up the situation by saying that the gene of Mendelian genetics has two separate identities: as a hypothetical material entity and as an instrumental entity. The future development of genetics was the result of the interplay between these roles. Griffiths and Stotz suggest a different reading: "Most recent scholars agree that the real achievement of classical Mendelian genetics was not a theory centered on a few principles of high generality, but rather an experimental tradition in which the practice of hybridising organisms and making inferences from patterns of inheritance was used to investigate a wide range of biological questions" (2013, p. 17). "Classical genetics was not a theory under test, or a theory that was simply applied to produce predictable results. It was a method of expanding biological knowledge" (2013, p. 19). Similar points have been raised by Darden (1991); Waters (2004, 2006); and Weber (2005).

This is clearly stated in Watson's (1913, p. 158) manifesto, "Psychology as the Behaviorist Views It," which opens with the following scathing remarks:

> Psychology as the behaviorist views it is a purely objective branch of natural science. Its theoretical goal is the prediction and control of behavior. Introspection forms no essential part of its methods, nor is the scientific value of its data dependent on the readiness with which they lend themselves to interpretation in terms of consciousness. The behaviorist, in his efforts to get a unitary scheme of animal response, recognizes no dividing line between man and brute. The behavior of man, with all of its refinement and complexity, forms only a part of the behaviorist's total scheme of investigation.

This programmatic presentation is an oversimplification, at best (cf. §3.4, Chapter 3). Various strands of mainstream behaviorism presuppose different methodological directives, objectives, and goals. Sure, by the end of World War I, the general redefinition of psychology as the scientific study of behavior had become widely accepted. Yet, Watson's uncompromising—and somewhat naïve—plan to reduce mentalistic psychology to neurophysiology was hardly the only or even the main player in the game. Tolman's "purposive behaviorism" and Hull's "mechanistic behaviorism" substantially differed from Watson's pioneering proposal. The same can be said for Skinner's "radical behaviorism," which is the object of our present focus.

Inspired by Watson, Skinner proposed a drastic overhaul of psychology by removing the causes of behavior from organisms and placing them in the outside environment. But, contrary to his predecessor, Skinner did not purport to reduce all mentalistic concepts to neurophysiological processes. In his academic maturity, Skinner acknowledged that a psychology dismissing all subjective, cognitive, and affective phenomena as meaningless would fall short of an adequate and comprehensive analysis of the human mind.

How did Skinner envision and enact his black-boxing of mental states? In what respects was he successful? What were his main shortcomings? This section addresses these questions by applying our three-step strategy.

Begin by focusing on the framing stage. Skinner's explanandum is the behavior of organisms. This, however, is a frame, a placeholder that stands in for a specific set of behavioral patterns, understood in a particular context that determines what should be explained and how. The context in question is specified in full by a model, the theory of radical behaviorism.

To appreciate the nuances and subtle complexity of the explanandum, consider what is being rejected. Prima facie, the answer seems obvious. As a behaviorist, Skinner had a strong aversion to traditional Cartesian characterizations and explanations of human conduct. Throughout his publications, he mentions various reasons for eschewing mentalism from psychology. So, what exactly is wrong with it? Answering this question, as Dennett (1981) notes, is not as simple as one may initially suppose. Skinner is undoubtedly familiar with classic philosophical problems with interactionism, fueled by the assumption that minds are non-physical substances. However, this worry is often downplayed. On a related note, Skinner is also aware that, by treating mental states as inherently private, Descartes places them out of reach for scientific inquiry. At the same time, cognizant of the insurmountable difficulties faced by Watson's naïve behaviorism, Skinner recognizes the need for science to posit some unobservable or unmeasurable properties. A third traditional objection to Cartesian dualism is that it considers mental states as "internal." While Skinner, like all behaviorists, finds this suspicious, in his better moments he sees that there is nothing obviously wrong with positing the existence of internal mediating states.

In sum, Skinner is aware of the long-standing philosophical qualms with Descartes's mentalism, and he is adamant in rejecting it. Nevertheless, pinpointing the decisive objections and, accordingly, the appropriate strategy to overcome them is no simple matter. This makes perfect sense from our perspective. The target is not behavior *simpliciter*. After all, virtually all psychologists are explaining behavior, Cartesian mentalists and Watsonian reductionists included. The real object of explanation is a set of behaviors framed by a model, the theory of radical behaviorism. Talk about "explaining behavior" is accurate. Still, we should recognize that it is a frame, a placeholder standing in for specific patterns of behavior to be covered.

With this in mind, let us move on to Skinner's causal explanations, which pertain to the second phase of his black-boxing strategy. From the standpoint of radical behaviorism, what are the difference-makers in the production of human conduct? Within Skinner's framework, the relevant set of behaviors is adequately accounted for when the investigator has identified all of the influences of which the behavior is a function. Consequently, the behavior itself is under experimental control. The antecedent influences acting on a behavior are the "independent" variables, whereas the behaviors which are a function of them are the "dependent" variables.

What exactly are these difference-makers? How should they be represented in the model? These questions bring us to the representation stage.

To appreciate the distinctive character of Skinner's proposal, it is useful to compare it with a different influential perspective, equally at odds with Cartesianism. In his article, "Critique of Psychoanalytic Concepts and Theories," Skinner (1956) contrasts his behaviorism with Freud's psychoanalytic approach. Skinner begins by praising Freud's general contribution to psychology, namely, "the application of the principle of cause and effect to human behavior. Freud demonstrated that many features of behavior hitherto unexplained—and often dismissed as hopelessly complex or obscure—could be shown to be the product of circumstances in the history of the individual" (1956, p. 77). From a behavioristic standpoint, however, Freud was guilty of a mistake. "Freud conceived of some realm of the mind, not necessarily having physical extent, but nevertheless capable of topographic description and of subdivision into regions of the conscious, co-conscious, and unconscious. Within this space, various mental events—ideas, wishes, memories, emotions, instinctive tendencies, and so on—interacted and combined in many complex ways" (1956, pp. 77–78). Freud failed to see, Skinner maintains, how this mental apparatus is a construction, which ends up being neither necessary nor helpful. The link between environment and behavior, he insists, is a direct one.

This point deserves elucidation. Following Flanagan (1991), let us represent a typical psychological causal chain in terms of three variables: environmental events (E), mental events (M), and behavioral events (B): $E \rightarrow M \rightarrow B$. To illustrate, I borrow an example from Leahey (2018, pp. 361–362). Consider a student who displays neurotic subservience to their teachers. A Cartesian will rationalize this behavior by appealing to the student's mental states. A Freudian would add an extra step. For instance, they might explain this by observing that, say, the student's father was a punitive perfectionist, who demanded unquestioned obedience. This image of a stern father has been internalized by the child, thereby affecting their behavior in the presence of authority figures. Radical behaviorism praises the Freudian move of recognizing how the conduct in question has an external cause. The key to explaining the student's actions lies in the relationship to their father. Skinner, however, takes this one step further by insisting that the link be direct. The inference to the unconscious image of an authoritative father in the superego explains nothing that cannot be explained better by reference to past behavior. The mention of this mental state adds nothing. Actually, it unnecessarily complicates the picture, as the nature of the mental state and its relation to environmental and behavioral variables must now be explained. And this is problematic. Mental states cannot be observed directly and are therefore underdetermined by evidence. This lack of depth is evidenced by Freud's metaphorical language. Speculative mythological stories depicting the mysteries of the mind are so enthralling that, regardless of their accuracy or plausibility, they end up stealing the show. This, Skinner argues, distracts from the difference-making factors that truly cause behavior and that can thus be used to effectively manipulate conduct. Teaching parents to raise well-functioning adults requires focusing on actual child-rearing practice, not fanciful tales about their mental life. In short, behavioral and environmental stimuli are the difference-makers responsible for behavior and, as such, they are the key to efficacious interventions.

We are now in a position to fully appreciate the radical behaviorist stance in action. Skinner's dismissal of mentalistic psychology, of both Cartesian and Freudian ilk, rests on a general methodological argument known as the *theoretician's dilemma*.[8] Return to our psychological causal chain of the $E \rightarrow M \rightarrow B$ form. What is the relation between these relata? Logic dictates that either mental states link environmental stimuli and conduct deterministically, or they must do so in a non-deterministic fashion. In the first case, Skinner argues, one can fully explain B without mentioning M at all. Internal mental states, including neurophysiological processes, turn out to be completely unnecessary for explaining behavior. Alternatively, if the $M \rightarrow B$ relation is indeterministic, then M itself becomes useless and thus, Skinner claims, it should be eliminated. Either way, references to mental states do not contribute to psychological explanations and should be omitted.[9] Environmental stimuli are all we need to account for behavior.

In sum, no mental properties mediate the relation between independent and dependent variables. References to consciousness and other superfluous mental states are eliminated once the controlling independent variables are properly understood. All mechanisms producing behavior are characterized in terms of environmental stimuli—E in our schema.

---

[8] The moniker "theoretician's dilemma" might sound puzzling to some readers. The rationale underlying this terminological choice is that, abiding to the conventions of logical empiricism, Skinner employs the expression "theoretical term" to refer to any unobservable object, law, or process. Thus, the name "theoreticians' dilemma" comes from its application to non-observable entities in science.

[9] As Flanagan (1991) notes, the argument, as stated, is neutral with respect to whether to eschew E or M. The fortified version of the dilemma thus runs as follows. One ought to avoid referring to mental events whenever possible because such references are logically eliminable *and* because they are epistemically problematic and useless in practice.

What is the nature of these difference-makers? How are they represented? Watson maintained that the causes of behavior should be expressed as physical processes. Skinner agrees that, in principle, physiology could re-describe these mechanisms. Yet, Skinner rejects Watson's naïve reductionism by making behavioral explanations autonomous from physiology.[10]

Mental and psychological states are Skinner's black boxes. Nonetheless, from his perspective, unpacking the frames and difference-makers posited by radical behaviorism is not an endeavor that will advance psychology.

Do Skinner's black boxes hold up to critical scrutiny? Does radical behaviorism provide an adequate account of human conduct and, specifically, of our mental capacities? My negative answer will likely not surprise many readers. Behaviorism has long been discarded in psychology, at least as a general theory of mind, and has been replaced by various cognitive approaches. The question that I want to raise is: what exactly was Skinner's capital sin?

Skinner's mistake, common to all forms of behaviorism, lies in his stubborn insistence on expunging mental states from psychology. Both his *pars destruens* and his *pars construens* turned out to be problematic. Beginning with the former, Skinner's general argument against appeals to mentality in psychology—the theoretician's dilemma—is fundamentally flawed. One horn of the dilemma assumes that, if links for the form $E \rightarrow M$ and $M \rightarrow B$ are deterministic, then appeals to M become utterly unnecessary. This is controversial, to say the least. But the other horn—if such links are indeterministic, appeals to M are useless—is completely off the mark. Non-deterministic generalizations can be perfectly explanatory, as witnessed by their constant and ubiquitous use across the sciences, from quantum physics to evolutionary biology, from neuroscience to economics.

Moving on to the constructive portion of the proposal, behavioristic explanations work reasonably well when applied to non-human animals, such as rats and pigeons, working in experimental laboratory settings. This is because, in these confined conditions, identifying the relevant stimuli and behaviors is relatively straightforward. Yet, once we apply this methodology to human beings "in the wild," identifying stimuli, behaviors, and their lawlike connections becomes much harder, arguably impossible, unless we posit a variety of logical, linguistic and, generally, cognitive competencies.[11]

---

[10] As Leahey (2018, p. 363) puts it, "Skinner assumed that physiology would ultimately be able to detail the physical mechanisms controlling behavior, but that analysis of behavior in terms of functional relationships among variables is completely independent of physiology. The functions will remain even when the underlying physiological mechanisms are understood."

[11] For related analyses, see Chomsky (1959); Dennett (1981); and Sober (2015).

Time to take stock. Behaviorists are often criticized for black-boxing mental states. Said critique is in need of clarification. Skinner does introduce various placeholders in the context of his causal explanations represented in a model. The relation between environmental stimuli, mental states, and conduct is described functionally. Contrary to Watson, Skinner refuses to reduce these psychological connections to physiology. These black boxes, frames and difference-makers alike, are no more mysterious or problematic than the ones constructed by Darwin, Mendel, and classical geneticists in biology. In this sense, Skinner's decision to black box is a success story, just like the ones discussed earlier. At the same time, Skinner's construction of black boxes dismisses the relevance of mental states in the production of behavior. The result—evident in all three phases of his construction: framing, difference-making, and representation—entails that his radical behaviorism leaves out an important aspect of psychology: mentality. This eventually caused his general model to collapse. This, in a nutshell, was Skinner's capital sin.

§6.5. Friedman's Black Boxes

At a general level of description, neoclassical economics can be viewed as an attempt to explain economic behavior by plugging choice data into the apparatus of rational choice theory. At the core of the approach lie the notions of *choice*, *preference*, and *utility*. To illustrate, suppose that an agent displays a consistent preference for *a* over *b* and for *b* over *c*. Neoclassical economics rationalizes and accounts for such behavior by positing that the agent attaches more utility to *a* than *b* and to *b* than *c*. In short, choice shows preference, preference reveals utility, and utility is what explains action.

As we saw in section 3.5 of Chapter 3, the revealed preference view of utility played an important role in turning economics into a testable science, conforming it to the strictures of logical positivism. Preference is displayed in choices, and choices are observable and, thereby, testable events. At the same time, this approach completely shields economics from the influence of any other inquiry. This is because, from the perspective under present scrutiny, choice and preference are not constructs open to further empirical analysis. They are "black boxes." This section analyzes the black-boxing of mental states enacted by neoclassical economists, by applying our three-step recipe.

To narrow the scope of the discussion, I focus on some examples famously developed by Friedman and Savage in their classic 1948 article, "The Utility

Analysis of Choices Involving Risk." Beginning with the framing stage, we can ask: what exactly is the object of their explanation?

Friedman and Savage's point of departure is a commonsensical observation. Human beings are constantly confronted with choices that differ, among other things, in the degree of risk to which they are subject. Should I accept a guaranteed or virtually guaranteed lower-salary job, or should I pursue a career that could lead to a higher income, but may also not work out? Is it wiser to invest my life savings in secure low-yield bonds or in more volatile stocks? Note that the problem is not restricted to economic decisions. These kinds of choices pervade virtually every aspect of our lives.

It is not uncommon for the same individual to respond very differently when confronted with distinct decisions along these lines. For instance, many of us enjoy gambling and also purchase insurance on our car or home. This raises an interesting question. Are these behaviors mutually consistent? At first blush, it might be hard for readers without an economic background to see why they would not be. Yet, consider the situation along the following lines. When I decide to insure my home, I am deliberately opting for the certainty of a small loss, the monthly premium that I pay to the insurance company, over the relatively remote possibility of a substantial loss, say, were the house to burn down. I know perfectly well that the "bet" I am taking is not a fair one. My monthly premium vastly exceeds the value of the house, minus my deductible, modulo the chance of something happening. This is precisely how insurance companies make profit. Still, I willingly eat this relatively minor cost to purchase my peace of mind. The situation is reversed when I buy a Powerball ticket, where the mirage of a huge gain is chosen over a sure-fire loss, namely, the cost of the ticket. Once again, there is no mystery that the bet is not fair. The lottery prize, no matter how sizable, is never commensurate to the expected value of the gamble in question. Yet, I agree to pay the premium for a tiny chance at great riches. Here is the rub. In the insurance case, I willingly pay a premium to choose certainty over uncertainty. In the gambling case, I pay a premium to choose uncertainty over certainty. In the former scenario, I am risk-loving; in the latter, I am risk-averse. Is it not inconsistent for the same individual to pay a premium to avoid risk, in one situation, and to pay a premium to seek risk, in the other? Is there any uniformity across these decisions?

Well into the twentieth century, the standard economic solution to this well-known puzzle involved drawing a distinction between two kinds of choices, "risky" and "risk-less" ones. The decision between certain options was explained in terms of maximization of utility. Economic agents were assumed to choose as they would if they attributed some common quantitative measure, "designated utility," to various goods. A rational subject then selects the bundle of goods that yields the largest total amount of utility. Choices under risk, in contrast, were explained in different terms, typically by appealing to ignorance of the odds or to personal dispositions of character.

Why was the same explanation—maximization of utility—not applied to both risk-less and risky choices? The reason is a direct consequence of the *principle of diminishing marginal utility*, which states that as an agent consumes more of a good, the marginal benefit of each additional unit of that same good decreases. It seems to follow from this basic economic tenet that no utility-maximizing agent would ever participate in a fair game of chance. This is because diminishing marginal utility entails that the gain from winning $+\$x$ will be less than the corresponding loss $-\$x$, rendering the utility of a fair bet negative. In short, the principle of diminishing marginal utility and the maximization of expected utility jointly imply that agents would have to be paid to offset the cost of bearing risk in a fair game of chance. The premium should increase to make an otherwise unfair bet appetizing. But this is plainly absurd. Myriad people constantly choose to partake in unfavorable gambles: lotteries, roulettes, risky investments, and the like. Thus, economists like Marshall rejected utility maximization as an analysis of risky choices. These widespread behaviors had to be explained in different terms.

Friedman and Savage's goal was to show how neoclassical economics could resolve this apparent tension. Contrary to a then widespread belief, risky and risk-less behavior could be subsumed under a unified explanation. The solution originates from Daniel Bernoulli's celebrated analysis of the St. Petersburg paradox, and it was molded into its modern form by von Neumann and Morgenstern's *Theory of Games and Economic Behavior*.[12] For the sake of brevity, in what follows, most technical details will be omitted from my reconstruction of Friedman and Savage's analysis. My goal is to break down their strategy in terms of our three black-boxing steps.

---

[12] The St. Petersburg paradox, also known as the "St. Petersburg lottery," is a puzzle related to probability and decision theory. A fair coin is flipped until it comes up heads the first time, at which point the player wins 2^n where n is the number of times the coin was flipped. How much should one be willing to pay for playing this game? The expected monetary value approaches infinity. And yet, it would be rejected by virtually any gambler. While the paradox was first invented by Nicolaus Bernoulli in 1713, it takes its name from its resolution by Daniel Bernoulli, Nicolaus's cousin, who at the time was a resident of the eponymous Russian city.

It should now be clear that the explanandum—why agent s chooses option a over option b, or vice versa—is far from an event *simpliciter*. The target in question is framed by rational choice theory, which lies at the heart of modern neoclassical economics. Some presuppositions of the model are stated clearly and explicitly. For instance, rational choice theory presupposes that, in making a decision, regardless of whether the alternatives are risky or certain, a consumer unit behaves as if the following were true. (i) The consumer unit has a consistent set of preferences. (ii) These preferences can be exhaustively described as a function attaching a numerical value ("utility") to alternatives. (iii) Finally, the objective of the consumer unit is to maximize expected utility, that is, to make its expected utility as large as possible.[13] Other aspects of the model are less explicit. For instance, what does it mean for the set of preferences attributed to a consumer unit to be "consistent"? Simply put, the requirement is that these preferences must not be contradictory and should not change over time. But how strictly should the temporal proviso be understood? On its strongest reading, an agent will never change its preferences. This, however, will make the model virtually inapplicable to real agents, as no human being displays identical preferences from the cradle to the grave. At the other extreme, if any whim can be accommodated as a change in taste, then the model becomes irrefutable, as consistency is the basic criterion for testing the hypothesis in question.[14] The solution to these issues cannot be stated explicitly. It must be determined pragmatically, on a case-by-case basis.

In short, when Friedman and Savage set out to provide a utility analysis of risky choices, terms such as "utility," "choice," and "preference" are used as frames. They are placeholders standing in for specific ranges of behaviors

---

[13] This, of course, is only one among many equivalent formulations of the axioms underlying the concept of rational choice presupposed in the neoclassical model. A broader, alternative characterization—which is closer to the original formalization provided by von Neumann and Morgenstern—states that an agent must choose in accordance with a system of preferences that satisfies the following properties. (a) The system is complete and consistent. (b) Any object which is a combination of other objects with stated probabilities is never preferred to every one of these other objects, nor is every one of them ever preferred to the combination. (c) If object a is preferred to object b, and b to object c, there will be some probability combination of a and c such that the individual is indifferent between it and b. This formulation shows more clearly that there is little difference between the plausibility of this hypothesis and the typical indifference-curve explanation of risk-less choices. Yet, it is more abstract than the one provided in the main text.

[14] Here is an example of behavior that would contradict the hypothesis. Imagine an individual who is willing to pay more for a gamble than the maximum amount she could win—for instance, a gambler who is willing to pay \$1 for a chance to win, at most, 99¢. Such an agent displays inconsistent preferences. Therefore, her behavior is economically "irrational" and, as such, it cannot be captured in terms of a monotonic utility function.

in need of explanatory coverage. Precisely which behaviors are subsumed under the analysis in question is explicitly stated or implicitly presupposed by the underlying theoretical model: the theory of neoclassical economics.

We can now focus on the causal explanation. The difference-maker for the framed explanandum—risky vs. risk-less choices—is the agent's assignment of utility to alternatives. As we saw in section 3.5 of Chapter 3, utility is a place-holder that plays a dual role. *Qua* frame, it stands in for patterns of behavior in need of explanation. *Qua* difference-maker, it stands in for mechanisms producing the behavior in question, thereby causally explaining it.

How should this difference-maker be represented? This leads us to the third and final phase of the black-boxing strategy. The explanations offered by neoclassical economics are explicitly advanced as "as if" claims. They do not presuppose that the agents themselves be able to specify the utility assigned to choices, or the precise shape of the utility function. As long as an agent behaves as if they are maximizing expected utility, the requirements of the framework have been met. Whether or not the agent can further provide an insightful rationale for their choices is interesting, but beside the point.

These framing assumptions have important implications for how utility is understood. Mainstream neoclassical economics depicts utility as a the-oretical construct, an ordinal ranking fully revealed by preference, with no cardinal counterpart. Utility, choice, preference, and other constructs are the black boxes of neoclassical economics. Friedman and his colleagues are ad-amant that any specification of the nature and structure of these boxes—for instance, how they are psychologically or neurally implemented by human agents—is of no consequence for economic theory itself.

Contemporary psycho-neural economics, in contrast, views utility as a neural value, instantiated in the brain and, therefore, quantitatively meas-urable, at least in principle. In both economic theories—neoclassical and psycho-neural economics—utility is a black box. Nevertheless, these boxes are structured very differently. In both cases, they are placeholders. But, except at very general levels of description (cf. §4.4, Chapter 4), these placeholders stand in for explananda framed differently and difference-makers distinc-tively represented. In short, they are quite different black boxes.

Does neoclassical economics provide an accurate and illuminating model of economic behavior? Does it capture the salient features of real-life eco-nomic decisions involving, for instance, investments, insurance, and gam-bling? These questions remain open. Beautifully simple mathematical precision, or drastically oversimplified mischaracterizations with little, if any, empirical import? Different readers will likely disagree on this judgment, based on their background and general methodological assumptions.

§6.6. Black Boxes in Science

Time to tie up some loose ends. A black box is a placeholder—frame or difference-maker—in a causal explanation represented in a model. This chapter has applied this general definition, first introduced in Chapter 5, to our case studies from the history of science. The tales of Darwin, Mendel, Skinner, and Friedman all have in common the deliberate omission of detail concerning central components of their explanations. These authors construct black boxes, whose basic structure is summarized in Table 6.1. Yet, significant differences persist.

Darwin successfully black-boxed the right difference makers: the mechanisms of variation and inheritance. Mendel identified the laws that govern inheritance and posited the existence of an entity—the gene or, in his words, the "factor"—that underlies this process. But, for Mendel and his followers early in the twentieth century, genes were still black boxes. These boxes were eventually opened, with the advent of molecular genetics, only to be subsequently replaced with other, more sophisticated black boxes.

Skinner purported to provide a comprehensive analysis of human conduct which dismisses the structure of the mind as causally irrelevant.

Table 6.1 Black Boxes in the History of Science

	Darwin	Mendel	Skinner	Friedman
Step 1: Framing	Evolution of species	Mendelian ratios	Psychological behavior	Economic behavior
Step 2: Difference-making	Natural selection	Mendel's Laws	Stimulus-Response	Maximization of utility
Step 3: Representation	Evolutionary theory	Classical genetics	Radical behaviorism	Neoclassical economics
Black Boxes	Competition Variation Fitness Inheritance (Descent with modification)	Mendelian factors Genes (Molecular genetics)	Mental and psychological states (Cognitive science)	Utility decision-making preference (Behavioral economics)

Unfortunately for him, mental properties do play a crucial role in the framing and production of action. His capital sin was not black-boxing per se, which is inevitable and important across the sciences. His mistakes pervade all three phases: framing wrong explananda, providing incomplete causal explanations, and representing them in inadequate models.

Are Friedman's black boxes more like Darwin's and Mendel's, or more like Skinner's? Is neoclassical economics the crown jewel of the social sciences or a regressive research project? This remains controversial, dividing economists and social scientists into opposing camps. I shall not attempt an answer here. I do, however, contend that the present framework provides the resources to recast, in more perspicuous and fruitful terms, the ongoing debate between neoclassical economists and psycho-neural economists. In doing so, it provides the foundation for more productive exchanges.

We are now in a position to fully comprehend, to a greater degree, the central claim advanced in Chapter 3. The history of science, I maintained, is a history of black boxes, constantly unwrapping old ones and constructing new ones. This process can be broken down to three constitutive steps: framing the explanandum, constructing a causal explanation, and representing these placeholders in a model. This chapter emphasized the prominent role of the three phases of black-boxing in the history of science. With all of this in mind, we can now move on to the philosophical payoff of all our hard work. Chapter 7 begins this final portion of the book with a critical discussion of the notion of *mechanism*, "black-boxing style."

7

Diet Mechanistic Philosophy

> Explanatory Knowledge opens up the black boxes of nature to reveal
> their inner workings.
>
> —Wesley Salmon, *Four Decades of Scientific*
> *Explanation*, p. 182

§7.1. Introduction

Chapter 5 broke down the construction of a black box into three constitutive steps. The framing stage sharpens the object of explanation. The difference-making stage provides a causal analysis of this explanandum by identifying the factors which influence significantly its production. The representation stage optimizes the explanatory "bang for the buck" by embedding the causal narrative into a suitable model. This led to the definition of black boxes as placeholders in causal explanations represented in models. I optimistically hope that readers find this intuitive enough. Still, this one-liner embeds various technical notions that required some elucidation.

Chapter 6 applied this three-step recipe to the case studies first introduced in Chapter 3. We saw that there may be various, often conflicting reasons for idealizing or abstracting away the structural details from a causal explanation. Sometimes, the underlying mechanisms are unknown, partially identified, or incorrectly understood. This situation was exemplified by the stories of Darwin and Mendel. The English evolutionist and the Bohemian geneticist were perfectly aware that there must be some biological apparatus governing the process of heredity. Still, whether or not they were personally invested in finding it, neither was able to do so successfully. In other circumstances, detail may be omitted because it is, rightly or wrongly, deemed irrelevant to the target at hand. This was the case with radical behaviorism. Skinner noted explicitly that treating organisms as black boxes is a drastic oversimplification. Nevertheless, he considered mental states and other "internal" psychological variables as disposable intermediaries that make little to no self-standing

Black Boxes. Marco J. Nathan, Oxford University Press. © Oxford University Press 2021.
DOI: 10.1093/oso/9780190095482.003.0007

contribution to human conduct. Finally, underlying entities and processes may also be dismissed as pertaining to an altogether different field of inquiry. This may also become an effective strategy for insulating a theory from "external" critique. Friedman and other neoclassical economists have long been aware that brain states provide the foundation for decision-making. Yet, a systematic investigation of their nature is bracketed as an endeavor for psychology and, as such, not a direct concern for economists working within this tradition. Setting these differences aside, our celebrated episodes have a common denominator. They all revolve around placeholders in causal explanations represented in models. This is to say, they all involve the construction of black boxes.

It is high time to address and enjoy the philosophical payoff of our hard work. To this end, the subsequent two chapters apply the black-boxing strategy to the concepts of emergence (Chapter 8) and scientific progress (Chapter 9). Chapter 10 aims to show that, by focusing on black boxes, one is able to embrace all the positives of both reductionism and antireductionism, while eschewing their more troublesome implications. Before doing so, the present chapter compares and contrasts the approach presented here with a movement that has gained much traction in the last couple of decades within the philosophy of science: the "new wave of mechanistic philosophy."

The new mechanistic philosophy, born as a reaction to the traditional reductionism vs. antireductionism divide, shares many core tenets underlying the strategy outlined here. This raises a concern. Is my treatment of black boxes as novel and original as I claim it to be? Or is it just a rehashing of ideas that have been widely discussed since the turn of the new millennium? As we shall see, the black-boxing recipe fits in quite well with the idea of science being in the business of discovering and modeling mechanisms. At the same time, the construction of black boxes, as I present it here, mitigates many of the ontological implications that characterize the contemporary landscape. For this reason, I provocatively refer to black-boxing as a "diet mechanistic philosophy" with all the epistemic flavor of traditional views, but hardly any metaphysical "calories."

Here is my plan for the pages to follow. Section 7.2 outlines the central tenets of the new mechanism. Section 7.3 emphasizes how the steps underlying the construction of a black box have been stressed, in some form or degree, in the contemporary literature on the discovery and representation of mechanisms. Still, these analogies should not be overstated. Section 7.4 stresses some differences and presents my constructive alternative. Section

7.5 draws a connection between mechanisms and black boxes. Section 7.6 discusses some implications of the "diet" approach, and section 7.7 wraps up the discussion with concluding remarks.

§7.2. The New Mechanistic Philosophy

The dawn of the new millennium has witnessed a revamped attention to the notion of *mechanism* across the philosophy of science. This claim calls for some clarification. Mechanism is hardly a novel concept. Natural philosophers began to characterize the world as a complex piece of machinery and, relatedly, to see science as fundamentally organized around the search for mechanisms at the outset of the scientific revolution, around the late sixteenth and early seventeenth centuries. The suggestive image of a mechanistic universe has been influential ever since. While these are important developments in the history of science, our focus in the present chapter is on a more recent and circumscribed phenomenon: the so-called new mechanistic philosophy.

The new mechanistic philosophy is a movement that digs its intellectual roots into the 1970s, when Stuart Kauffman and Bill Wimsatt pioneered an analysis of reduction in terms of breaking down wholes into parts. This decompositional strategy was subsequently developed by many others, including William Bechtel, Robert Richardson, Peter Machamer, Lindley Darden, and Carl Craver. Over the last couple of decades, the product of these conversations has spawned into a rich and lively research program.[1]

What is this movement all about? According to Glennan's (2017, p. 1) recent presentation, the new mechanistic philosophy "says of nature that most or all the phenomena found in nature depend on mechanisms— collections of entities whose activities and interactions, suitably organized, are responsible for these phenomena. It says of science that its chief business is the construction of models that describe, predict, and explain these mechanism-dependent phenomena." From this perspective, most—arguably all—phenomena found in nature depend on mechanisms. What are these

---

[1] Without any pretense of comprehensiveness, my succinct overview borrows from a selected handful of representative works. In particular, the following discussion draws heavily from three recent monographs: Bechtel and Richardson (2010); Craver and Darden (2013); and Glennan (2017). Incidentally, to the best of my knowledge, the expression "new mechanistic philosophy" was coined by Skipper and Millstein (2005).

mechanisms? Unfortunately, a clear, precise, and widely accepted definition is still wanting. Nonetheless, as the preceding quote suggests, a mechanism can be broadly characterized as an organized aggregation of entities and activities that collectively produces a specific phenomenon.[2]

Thus presented, one could legitimately wonder, what makes this mechanistic movement novel, substantive, or even interestingly controversial? Who would contest the preceding pronouncements? Would anyone seriously disagree with the statement that the natural world is replete with organized collections of entities and activities that produce phenomena, some of which are explanantia or explananda of scientific interest?

Clearly, there must be more to it. And, indeed, there is. The main philosophical contribution associated with the new mechanism, its signature move, is a change in focus, namely, a shift away from the primacy of laws and theories toward a renewed attention to mechanisms and models. This claim calls for further elucidation. From the perspective of most twentieth-century philosophy, science is first and foremost a search for natural laws. In general, concrete instances of events, together with the mechanisms that produce them, were only interesting to the extent that they provide evidence—confirmation or disconfirmation—of these generalizations. The particularist ontology underlying the neo-mechanistic approach purports to turn this relation on its head. Generalizations have heuristic value. But they hardly mirror reality, true states of affairs. What exist are mechanisms, which can be discovered, studied, and represented in scientific models. The implication, of course, is not to deny the existence of laws, or law-like generalizations, which can be quite useful in predicting and explaining what happens in the world. The point is rather that laws are less fundamental than they were taken to be. Such generalizations are produced, sustained, and explained by mechanisms. It is the existence of a mechanism that grounds a law, not the other way around. In this sense, laws, like much else that goes on in science, are mechanism-dependent: descriptions of the behavior produced by mechanisms. These theses are more substantial than the preceding truisms.

Equipped with these insights, the mechanistic movement aims to provide a new framework in which to tackle a number of classic problems in

---

[2] While this is undoubtedly an oversimplification, consider: "A mechanism for a phenomenon consists of entities (or parts) whose activities and interactions are organized so as to be responsible for the phenomenon" (Glennan 2017, p. 17). "Mechanisms are entities and activities organized such that they are productive of regular changes from start or set-up to finish or terminating conditions" (Craver and Darden 2013, p. 15, italics omitted).

the philosophy of science. Central among them is the nature of explanation, where the emphasis on mechanisms and derivative notions constitutes an effective antidote to the outmoded deductive-nomological approach. But the promises of the new mechanism are hardly confined to the explanatory enterprise. The philosophy of science in the new millennium is replete with mechanism-based accounts of causation, reduction, models, reasoning in discovery, underdetermination, laws, theories, and other classic concepts.

How successful are these attempts? This remains an open question. Sure, the new mechanistic philosophy has gained a large—and still growing—number of adepts. At the same time, in addition to praise and support, the new mechanism has also triggered objections, ranging from unimpressed skeptics to outright critics. I defer a discussion of naysayers to section 7.7, which sketches a black-boxing-oriented response to counterarguments. Before addressing the negatives, let us focus on the positives. Section 7.3 discusses some remarkable analogies between mechanistic approaches and black-boxing.

§7.3. Three Steps: A Neo-Mechanistic Perspective

The term "mechanism" has been used extensively throughout this book. Recall our distinction between two types of placeholders. Frames stand in for behaviors in need of explanation. Difference-makers stand in for the mechanisms that produce the behaviors in question. Such use of "mechanism" falls perfectly in line with the definition presented in section 7.2: a collection of entities and activities generating a specific phenomenon. Indeed, as we shall now see, the analogies between the new wave of mechanistic philosophy and black-boxing cut much deeper than this shared minimal definition.

In their recent monograph, Craver and Darden (2013) provide an influential analysis which breaks down the discovery of mechanisms into four main phases. First, one must characterize the phenomenon under investigation. This involves providing a precise-enough description of the behavior in need of explanation. The second step involves representing the mechanism which produces the behavior in question. This is done by constructing a schema which generates a space of possible mechanisms responsible for the explanandum. Third comes the evaluation of this schema. In the fourth revisitation phase, the initial representation can be amended and improved.

This process should have a familiar ring to it. Craver and Darden's four stages resemble the three steps of the strategy detailed in Chapter 5. These similarities are hardly haphazard. They are the consequence of a shared scientific outlook. Indeed, black-boxing echoes many important insights stressed by the new wave of mechanistic philosophers, beginning with its broadly naturalistic stance. This section shows how all three steps inherent to the construction of black boxes—framing, difference-making, and representation—are present, in some form or degree, in the neo-mechanistic literature and, more generally, across contemporary philosophy of science.

Does this mean that black-boxing is just the repackaging of old ideas and well-known adages? No, it does not. Section 7.4 will stress some key points where the present account departs, quite drastically, from traditional mechanistic approaches. Before doing so, however, we should focus on similarities.

Begin with the first stage of the black-boxing strategy. Recall from section 5.2 of Chapter 5 that framing involves specifying and contextualizing the object of explanation. Making the explanandum fully explicit requires the construction of a model, which is no trivial task, especially in the context of a scientific hypothesis. Fortunately, in order to get the inquiry going, it is sufficient to specify a frame, a preliminary scaffolding that provides a coarse-grained depiction of a range of behaviors in need of explanation.

Framing is the preliminary step in model-construction. Craver and Darden (2013, p. 62) posit an analogous process in the building of mechanistic models:

> The search for mechanisms must begin with at least a rough idea of the phenomenon that the mechanism explains. A complete characterization of a phenomenon details its precipitating, inhibiting, and modulating conditions, as well as noting nonstandard conditions in which it can (be made to) occur and any of its by-products. During discovery episodes, a purported phenomenon might be recharacterized or discarded entirely as one learns more about the underlying mechanisms. Lumping and splitting are two common ways of revising one's characterization of the phenomenon in the search of mechanisms.

The issue here is not merely the truism that, in order to discover the mechanism underlying a certain kind of behavior, one must first determine which behavior is under investigation. The contention is stronger. Mechanists are adamant in stressing that all mechanisms are mechanisms *for* some phenomenon. This entails that the very nature and boundaries of a mechanism depend, in large part, on the phenomenon being produced.[3]

In short, the inherently perspectival identity of mechanisms and the explanatory relativity underlying the black-boxing strategy point to the same core phenomenon. The significance of a framing process lies in sharpening the explanandum and determining the nature and boundaries of the explanans.

Moving on to the second stage of black-boxing, difference-making provides a causal explanation of the framed explanandum by identifying the factors that make a significant contribution to its occurrence. Once again, I do not have a novel account of causal explanation to offer. Quite frankly, I am skeptical that that any single, monolithic definition can be nuanced and general enough to subsume the broad range of phenomena classified as causes. Still, black-boxing, as presented here, need not commit to any specific approach to causal explanation. Any theory that specifies difference-makers of an event may be employed in the construction of a black box.

These observations raise questions. How should we understand the relation between mechanisms and causation? In what ways is the black-boxing recipe similar to and different from the new wave of mechanistic philosophy?

The exact connection between mechanisms and causation is a controversial matter. In general, three categories of approaches can be distinguished. First, there are authors who argue that mechanisms are the key for understanding the nature of causation. For instance, Machamer (2004), Bogen (2008), and Glennan (2017) have articulated theories of causation, all of which take mechanisms to be truth-makers for causal claims. A second family of views encompasses various responses which reject the need for a distinctively "mechanistic" approach to causation. Craver's (2007) analysis of explanation in neuroscience, for one, borrows from manipulability theory, a non-reductive approach falling within the difference-making camp.

---

[3] Glennan (2017, p. 44) is explicit on this point: "The fundamental point is that boundary drawing—whether spatial boundaries between parts of mechanisms or between a mechanism and its environment, or temporal boundaries between the start and endpoints of an activity or mechanical process—has an ineliminable perspectival element. But the perspectives from which these boundaries are drawn are not arbitrary or unconstrained. The perspective is given by identifying some phenomenon. This phenomenon is a real and mind-independent feature of the world, and there are real and independent boundaries to be found in the entities and activities that constitute the mechanism responsible for that phenomenon." Similarly, Craver and Darden (2013, p. 52) stress that "characterizing the phenomenon to be explained is a vital step in the discovery of mechanisms. Characterizing the phenomenon prunes the space of possible mechanisms (because the mechanism must explain the phenomenon) and loosely guides the construction of this hypothesis space (because certain phenomena are suggestive of possible mechanisms)."

Third, and finally, there are philosophers, such as Bechtel and Richardson (2010) and many of their collaborators, who view mechanisms primarily as epistemic and explanatory constructs. Consequently, they try to avoid altogether the thorny ontological issue of causation. In short, there is no official consensus among mechanists regarding the relation between causes and mechanisms. This remains an important open issue.

It should be evident that the black-boxing strategy fits quite naturally with the second and third paths just delineated. Any account of difference-making can be squared with the story sketched in section 5.4 of Chapter 5 and, if mechanisms have nothing to say about causation, we can pick our favorite alternative. This, however, is not to say that black-boxing is flatly at odds with mechanistic or process-based accounts of causation. Causal relevance—the heart and soul of difference-making—is such a central concept that all theories of causation and causal explanation, mechanistic or otherwise, must say something about it. After all, can we really do science without difference-makers or some surrogate?

In sum, the relation between mechanisms and causation is complicated. Some contend that mechanistic philosophy has something to say about the nature of causation. Others rest content with supplementing mechanistic insights with preexisting concepts. Either way, any theory of causation and causal explanation worth its salt requires a notion of relevance. If my story of difference-making is not quite it, it must be something rather similar.

Finally, let us consider the third and final phase of black-boxing. The representation stage seeks to embed the causal narrative into a suitable model. This is neither original nor controversial. The importance of models and other vehicles of representation is hardly novel. As mentioned in section 5.4, the idea that models function as mediators between theories and the world has become rather commonplace, at least among philosophers of science.

The new wave of mechanistic philosophy has incorporated and developed these insights by focusing—unsurprisingly—on the representation of mechanisms. Mechanistic models describe entities and activities responsible for specific behaviors. A distinction is commonly drawn between two components. The "phenomenon description" provides a model of the explanandum; the "mechanism description" depicts the system that produces the behavior in question. Once again, this dichotomy should have a familiar ring to it. It roughly corresponds to the two kinds of placeholders introduced in Chapter 4. Phenomena descriptions are what I called "frames." Mechanism descriptions pick out difference-makers that bring about the explanandum.

What about the processes of abstraction and idealization required to optimize the explanatory power of mechanistic representations? This, too, has been extensively debated in the neo-mechanistic literature, generating a variety of nuanced positions, which can be placed on a spectrum.

Some authors have acknowledged how abstracted and idealized depictions inevitably conflict with the ideal of completeness, without affecting their overall efficacy. Bechtel (2011) and Glennan (2017), for instance, stress that even the most complete mechanistic models require a certain degree of abstraction, in order to isolate their targets. Thus, even models purporting to represent the details of actual mechanisms will typically be highly idealized and unrealistic, in significant respects. The gradual transition toward more and more complete models is a regulative ideal, which involves inevitable trade-offs reflecting, in part, technological and cognitive limitations and, in part, the diversity of explanatory models. From this standpoint, accuracy is an important goal, albeit one that is typically bought at the expense of other theoretical virtues, such as completeness and generality.

Other mechanistic philosophers, who tend to view completeness as the chief explanatory ideal of the mechanistic sciences, are less enthusiastic about omission of detail. Initially, the prominence of completeness over other theoretical virtues was formulated quite crudely (Craver 2007). This led critics to interpret it as a problematic "the more details a mechanistic representation contains, the better" stance. More recently, Craver and his collaborators clarified their view, which should be understood as claiming that the more *relevant* details, the better (Craver and Kaplan 2020). Evidently, describing *all* the entities, activities, and organizational features of a mechanism would yield a hefty depiction, unwieldy for purposes such as prediction and control. Hence, some degree of abstraction is required to make descriptions perspicuous and parsimonious, relative to their explanatory target. But this is a necessary evil, not a virtue of our models.

As we shall see in section 7.5, the mechanistic approach to abstraction and idealization—and, especially, Craver's "3M" approach—is different from my black-boxing strategy in significant respects. Setting these nuances aside, the important point, for the time being, is simply the role of abstraction and idealization in making models congruent with the portion of the world they represent and explain. This core aspect of my third representation stage has also been recognized and stressed in the mechanistic literature.

Time to take stock. All three constitutive phases of black-boxing—framing, difference-making, and representation—figure, in some form or degree, within the new mechanistic philosophy. Once again, these considerations raise a nagging concern. Is black-boxing no more than a repackaged version of the new mechanism? The short answer is: No. Analogies, unquestionably robust, should not be overstated. To this point, the remainder of the chapter emphasizes some crucial discrepancies. These differentiate the construction of black boxes from mainstream versions of mechanism.

§7.4. Mechanisms, Models, and Metaphysics

In a nutshell, the distinctive feature of the notion of mechanism underlying my analysis of black boxes is its relativization to models and, consequently, its lack of ontological import. The present section presents this context-relativity and discusses some of its implications.

Let us begin with a truism. There is a distinction between entities in the world and their representations. No scientist or philosopher worth their salt denies this, with the possible exception of idealists—advocates of a radical position that I set to the side. Mechanists are no exception. It is common within the movement to talk about *mechanisms* as things "out there," on the one hand, and the *mechanistic models* that depict them, on the other.

Nevertheless, potential confusion is just around the corner. The root problem is simple. The term "mechanism" is ambiguous. It may refer to things in the world or to their depiction. To wit, consider the following situation, admittedly oversimplified, but not altogether unrealistic. Suppose that a team of researchers is investigating the process of gene expression in eukaryotic cells. In the course of the project, they construct a model of the phenomenon under investigation, like the one depicted in Figure 7.1, very much akin to the ones found in elementary cytology textbooks. This raises a concern. When we talk about, say, the "mechanism" that unwinds DNA, are we talking about actual cellular processes or their representations?

This question will undoubtedly raise more than a few eyebrows. Asking whether we are talking about cells *or* representations is foolhardy. Obviously, we are talking about *both*. Models, mechanistic or otherwise, represent entities and processes in the world. Therefore, claims about representations are, *ipso facto*, claims about the world. To get the point across, consider the

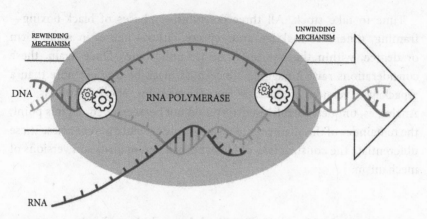

Figure 7.1. A simple model of gene expression.

following analogy. Imagine holding a picture of the Mont Blanc and uttering, "That mountain is beautiful!" Am I talking about the picture or the mountain itself? Obviously, since the photograph is supposed to faithfully mirror reality, I am referring to both: the mountain depicted in the photo is beautiful. The same applies to genes and cells. When I claim that gene transcription requires the unwinding of the DNA double helix, this claim holds in the model *and* in reality. In short, the question whether "mechanism" refers to models or reality is wrongheaded. It refers to both.

Intuitive as it may seem, this response is overly simplistic. Let me be very clear that I do not intend to deny that models purport to represent bits of the world. Of course they do. Similarly, I know better than to question the existence of a physical reality—idealism, as noted earlier, is out of the equation. My point is rather that the model-world relation is much more nuanced, complex, and indirect than the photograph analogy would suggest. And, as we shall see, this has robust philosophical implications.

To begin, take a closer look at the diagram in Figure 7.1. This representation of the cellular environment is inaccurate in several respects. First, DNA is depicted as a self-standing double helix in which all base pairs are available and accessible to protein. Second, all enzymes and transcription factors are either absent or tacitly assumed to be present in just the right quantity. Third, there are no repression mechanisms, such as DNA methylation, that could potentially interfere with gene transcription. Finally, the cellular environment is represented as a uniform, spacious system in which molecules are free to roam aimlessly without impediments.

In short, this simplified model represents a set of ideal conditions that are seldom or never instantiated in reality.[4] Real-life cells, it goes without saying, are way more complicated. DNA is tightly coiled around histones in ways that prevent enzymes from accessing regulatory or structural regions of the gene. Methyl-groups hide nucleotides from transcription factors. Proteins required for the expression of the gene are often absent or inaccessible. And, finally, molecular interactions do not occur in a void but, rather, in a crowded ecosystem. These details are not minor omissions or irrelevant by-products. They are what turn formless blobs of cells into finely tuned organisms capable of selectively coordinating the transcription and translation of numerous genes in the right place and at the right time.

These considerations invite a natural question. Why would biology textbooks deliberately provide simplified representations that do not accurately represent real organisms? The obvious answer is that biological organisms are extremely complex. Describing all the components of an actual cell would complicate the representation exponentially. The oversimplification, from this standpoint, is a small price to pay in exchange for tractability.

There is surely something true captured by this intuitive answer. Pragmatic convenience is definitely a factor. But is this all there is to it? The answer is negative. To see why, it is crucial to draw a distinction between *abstraction*, the omission of detail, and *idealization*, the deliberate misrepresentation of detail. Tractability may explain the presence of abstraction in the model. But how do we account for the deliberate distortion in the exact places where the mechanistic descriptions gain force? Why present a model that, as mentioned earlier, captures neither necessary nor sufficient conditions for gene expression?[5] It is one thing to leave things out. But why the flat misrepresentation of reality? In short, perspicuousness is only half of the story, and not the controversial part. So, the issue remains: why would textbooks intentionally provide inaccurate representations?

---

[4] More precisely, whether these ideal conditions are "seldom" or "never" instantiated depends on how strictly we interpret the representation itself (Nathan 2015). On a looser reading, while real-life cells sometimes instantiate the conditions portrayed in the diagram, they often do not. From a stricter perspective, one could say that the conditions depicted here are never realized. However, some real-life conditions resemble the situation more closely than others. For our purposes, we need not choose between these two alternatives. The important point, from the present standpoint, is that real cellular environments are substantially different from the diagrammed ones which, *sensu stricto*, fail to provide necessary or sufficient conditions for the expression of genes.

[5] For a detailed discussion of how mechanistic explanations often intentionally distort the central difference-makers in causal explanations, see Love and Nathan (2015).

A better answer becomes available as soon as we realize that the goal of the model is not to accurately *describe* cells. What the diagram purports to do—and does so quite well—is provide an *explanation* of gene expression, by showing how the underlying processes are instantiated in abstract, idealized settings. When the actual cellular environment suitably resembles or approximates the conditions depicted in the diagram, genes are, in fact reliably transcribed. In this respect, our model is analogous to the ideal gas law ($PV = nRT$) or to the identification of water and H_2O. These, too, are idealized conditions that are seldom or never instantiated. No real substance satisfies the ideal gas law, and pure H_2O is extremely rare outside the lab. Distorted as they are, these depictions are nonetheless highly explanatory.

The thesis that describing and explaining are goals that, generally, do not go hand in hand has been articulated systematically by Nancy Cartwright. In an influential article aptly entitled, "The Truth Does Not Explain Much" (1980), she argues that many philosophers are guilty of conflating two functions of science that should be kept distinct: *truth* and *explanation*. More specifically, explanation is commonly perceived as a by-product of truth. Scientific theories are taken to explain by dint of the description they give of reality. It follows that descriptive accuracy is all that science is really after. Explanatory power is a welcome free lunch.

This misconception, Cartwright goes on to argue, was fostered by the deductive-nomological model and its presupposition that all explanation requires is knowledge of laws of nature, in addition to some basic logic and probability theory. This is a mistake, and a pernicious one too. There are perfectly good explanations that are not covered by any law. In addition, and more provocatively, read literally, these laws are not only false; they are *known* to be false. Tacitly prefixed with *ceteris paribus* hedges, they only cover those sporadic cases where the conditions are just right. Either way, this is bad news for covering-law approaches to explanation.

Forty years after the publication of Cartwright's essay, the deductive-nomological model is all but dead and gone. Few philosophers still believe that explanation can be fully reduced to the logical derivation of an explanandum from a complex explanans that contains laws of nature together with a specification of initial conditions. Yet, interestingly, the identification of description and explanation into a single indivisible goal lives on in the new wave of mechanistic philosophy.[6] Many contemporary philosophers suggest that scientific explanation is ultimately a matter of describing mechanisms. As noted in section 7.3, whether such explanations should be as accurate as possible remains an open and controversial matter. Still, there is a widely shared presupposition that mechanistic descriptions take care of both endeavors at once.[7] A single model, sufficiently rich in detail, can simultaneously provide accurate descriptions and illuminating explanations.

What does any of this imply concerning the representation of mechanisms? The relativization of theories, models, and inquiries points to the importance of a different and less intuitive kind of idealization, namely, *multiple-models idealization*. This is the practice of constructing multiple, related, but incompatible models that capture distinct and independent aspects of a complex system.[8] The idea of multiple, independent models bears significant metaphysical implications. First, explanatory relativity threatens the ontological import of explanation. Over the last few decades, it has become fairly popular among philosophers to adopt an "ontic" conception of explanation. From this perspective, "explanations are objective features of the world" (Craver 2007, p. 27). The necessary construction of various incompatible frameworks, inherent to multiple-models idealization, is hard to reconcile with this kind of objectivity. Explanation is not a matter of ontology. It is ultimately an epistemic act, a matter of representation.[9] Some mechanists have conceded this point and have responded accordingly. For instance, Glennan (2017, Ch. 3) endorses the importance of abstraction and idealization and

[6] This has a paradoxical flavor, given the instrumental role of mechanists in replacing deductive-nomological approaches in favor of more naturalistic analyses of explanation.

[7] For instance, Glennan (2017, Ch. 3) criticizes attempts to distinguish between "phenomenal" and "mechanistic" models. He does so on the grounds that separating these functions requires an appeal to the intentions of the modeler.

[8] The concept of multiple-model idealization is borrowed from Weisberg's (2013, Ch. 6) three kinds of idealization. In addition to the multiple-models approach, Weisberg adds two more variants. The first is *Galileian idealization*, the practice of introducing distortions for the sake of simplifying theories. The second, *minimalist idealization*, corresponds to the practice of introducing only causal factors that make a difference to the occurrence of an event or phenomenon. All three kinds of idealization play a prominent role within the black-boxing strategy. Multiple-models follow from the framing stage. If models are relativized to the framing of an explanandum, then reframing the object of the explanation along alternative lines will require a different model, with varying abstractions and idealizations. Minimalist idealization occurs at the difference-making stage, when the causes that significantly influence the (framed) explanandum are distilled. Galileian idealization chiefly pertains to the representation stage, when the causal explanation is embedded in an appropriate model, that should be presented as perspicuously as possible.

[9] This point was originally developed in Love and Nathan (2015). For a more overarching critique of the ontic conception of explanation, see Wright and Van Eck (2018).

explicitly rejects completeness as a mechanistic ideal. Similarly, Bechtel and Richardson (2010) eschew ontic approaches to explanation in favor of representational, epistemic views.

But the importance of multiple incompatible models triggers even more radical implications when coupled with the explanatory relativity discussed in section 5.2 of Chapter 5. Specifically, it suggests that a single notion of mechanism cannot simultaneously play an ontological and an epistemological role. Allow me to elaborate. At the beginning of this section, I noted that the distinction between mechanisms and mechanistic models is both commonplace and uncontroversial. Still, it is typical to freely employ the term "mechanism" to refer to both entities *and* their representations. This might appear like a natural move: after all, like photographs, models purport to represent real-world systems. Yet, this is problematic, as the idealized entities postulated in models are quite different from the complex universe we inhabit. In short, real mechanisms are typically nothing like represented mechanisms. Consequently, extrapolating properties from one to the other requires more than a bit of caution. Employing the same term—"mechanism"—to refer to real and idealized entities can only foster confusion. For this reason, I propose the following terminological convention. I shall use the terms "process" and "system" to refer to actual entities in the world. "Mechanism" is used exclusively to pick out model-theoretic representations of these systems.

In conclusion, it is of paramount importance to distinguish two aspects of the practice of introducing and modeling mechanisms: ontology and epistemology. This point is stressed clearly by Glennan (2016, p, 796):

Philosophers have contemplated mechanisms in connection with two different kinds of projects. The first is broadly metaphysical, investigating the nature of mechanisms as an ontological category and exploring how mechanisms may be related to or explain other ontological categories like laws, causes, objects, processes, properties, and levels of organization. The second is methodological and epistemological, exploring how mechanistic concepts and representations are used to explore, describe, and explain natural phenomena. Opinions are divided as to how these two projects are related. Some philosophers see mechanisms as part of the deep structure of the world and ground the methodological importance of mechanistic thinking in this fact. Others have a more instrumentalist view in which mechanistic thinking is an important part of science but is one that is deeply pragmatically tinged.

On this realism vs. instrumentalism spectrum, I fall decisively in the latter camp. "Mechanism" here becomes a pragmatic vehicle of representation, devoid of virtually any ontological import. With these considerations in mind, section 7.5 brings black boxes back into the equation. Section 7.6 presents the advantages of this "diet" approach.

§7.5. Mechanisms and Black Boxes

The previous section (§7.4) attempted to draw a wedge between the metaphysics and the epistemology of mechanisms. Specifically, I argued that mechanisms should not be characterized as ontological posits. Mechanisms are not things "out there" in the world. Rather, mechanisms are best understood as epistemic constructs, model-theoretic vehicles of representation. Section 7.6 will argue that, by adopting this deflationary stance, one may effectively respond to some concerns that have been raised against traditional mechanistic theory. Before doing so, the present section discusses the connection between mechanisms and black boxes. After clarifying the nature of the relation in question, I compare and contrast my conception of black boxes with two influential accounts taken from the two most systematic accounts of black boxes in the philosophical literature, to the best of my knowledge.

The upshot of our discussion is that mechanisms are best understood as vehicles of representation. Given that models may depict real systems at different scales, it seems reasonable to insist that mechanisms are a kind of placeholder. After all, a depiction, verbal or pictorial, "stands in" for the system it purports to represent. All of this is, hopefully, rather intuitive. Attentive readers, however, will surely note that this claim is in tension with some tenets advanced in the first half of this book. Let me explain.

Back in Chapter 5, black boxes were defined as placeholders in causal explanations represented in models. Previously, Chapter 4 distinguished two kinds of placeholders: frames, which stand in for patterns of behaviors in need of explanation, and difference-makers, which stand in for the mechanisms that produce the behavior in question. It follows that one kind of black box—namely, difference-makers—are, for all intents and purposes, placeholders standing in for mechanisms, framed and represented in a model. The problem should now be evident. It looks like a mechanism is a black box, and a black box stands in for a mechanism. But this seemingly makes no sense: a mechanism cannot stand in for itself! What is going on?

My proposed solution is simple and, hopefully, intuitive. There is no paradox here: mechanisms are black boxes and black boxes stand in for mechanisms. The key lies in relativizing these statements to different levels of description and explanation. To illustrate, consider the scenario depicted in Figure 7.2. We have a mechanism (M_n) that is described at a specific level (L_n). Now imagine providing a coarser description (M_{n+1}) of mechanism M_n at higher-level L_{n+1}. Note that M_{n+1} contains a black box. This black box is a placeholder that stands in for mechanism M_n. Thus, at this coarser scale, the mechanism described in M_n is a black box. The process also works in the opposite direction. M_n itself will contain a number of black boxes. These black boxes stand in for the mechanisms that are left unpacked in M_n. We could open and, thereby, eliminate these black boxes by providing a finer-grained description of mechanism M_{n-1} at a lower level L_{n-1}. Doing so would replace them with a series of mechanistic descriptions which, in turn, would themselves contain a series of black boxes. This process may continue indefinitely, until we reach an ontological bedrock of entities, if such ground level even exists. Otherwise, it's black boxes—like the famous turtles—all the way down!

In sum, a mechanism can be defined as a model-theoretic representation of an organized aggregation of entities and activities that collectively produce a specific phenomenon. A black box—more precisely, the type of black

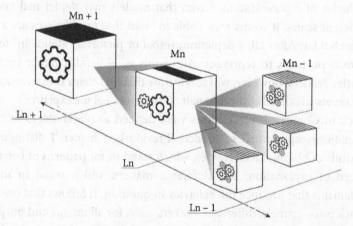

Figure 7.2. Mechanisms and black boxes are relative to levels of explanation.

box that I have dubbed a "difference-maker"—is also a representation of a causal relation relative to a certain scale. Both mechanisms and black boxes are placeholders: they stand in for more comprehensive descriptions that produce an effect. But there is no tension between the claims that a mechanism is a black box and that a black box stands in for a mechanism. The reason is that the same model-theoretic depiction counts as a mechanism at one level of description, and as a black box, relative to a different, coarser grain. These considerations raise important questions concerning the relation between descriptions at different levels of the hierarchy:

$$\ldots L_{n-1}, \ldots L_n, \ldots L_{n+1}, \ldots$$

These issues will be addressed in Chapter 10. In the meantime, in the remainder of this section, I further clarify my perspective on the relation between mechanisms and black boxes by comparing and contrasting it with two influential depictions due, respectively, to Craver and Strevens.

In a classic discussion, Hanson (1963, Ch. 2) draws a distinction between three different kinds of "boxes" in science: black, gray, and glass. Simply put, a black box is the representation of a phenomenon where the component functions are unknown. A gray box is a model where all components are specified in some degree of detail. Finally, a glass box is a depiction of the target system that is complete or, more modestly, complete for the purposes at hand. All relevant components are exhaustively depicted. Elaborating on this distinction, Craver characterizes the process of scientific discovery as the gradual transformation of a black box into a gray box and, eventually, into a glass box.[10] This progressive opening of boxes is part of the general process of turning a "mechanistic sketch"—an incomplete depiction that characterizes parts, activities, and features of a mechanism while leaving gaps—into a "complete mechanistic model," that is, a comprehensive description of all the components and activities of a mechanism. From this standpoint, a black box is part of a mechanistic sketch. It is something that marks a "hole" in our knowledge of how things work. This, Craver notes, does not necessarily invalidate the description in question. Ignorance is not a crime: just ask Darwin! Yet, these gaps are often masked by filler terms that create a misleading impression that the explanation in question is complete.

---

[10] See Craver (2007) and Craver and Darden (2013).

Black boxes, question marks, and acknowledged filler terms are innocuous when they stand as place-holders for future work or when it is possible to replace the filler term with some stock-in-trade property, entity, activity, or mechanism (as is the case for "coding" in DNA). In contrast, filler terms are barriers to progress when they veil failures of understanding. If the term "encode" is used to stand for "some-process-we-know-not-what," and if the provisional status of that term is forgotten, then one has only an illusion of understanding. For this reason, neuroscientists often denigrate the authors of black-box models as "diagram makers" or "boxologists." (Craver 2007, pp. 113–114)

As I understand him, Craver draws attention to two important points. First, the construction of black boxes is a necessary phase in science. When dealing with complex systems, such as the human brain, many constituents, entities, and activities are bound to be unknown. Nevertheless, explanation can proceed steadily in the face of ignorance and error by black-boxing the unknown processes, or parts thereof. These coarse-grained functional analyses containing black boxes can be viewed as "mechanism sketches".[11] A good illustration of this is provided by Darwin's and Mendel's treatment of the mechanisms of inheritance. Second, there is nothing wrong with black-boxing when the black box explicitly signals our lack of knowledge. Yet, over time, these filler terms become increasingly crystallized and ingrained in our practice. Consequently, their tentative, provisional status tends to be forgotten. When this happens—when a black box is misleadingly presented as a gray box or as a glass box—we only have an illusion of understanding, not true comprehension.

I agree with both remarks, which also apply to my own account. At the same time, it is important to note that the notion of black boxes developed in the previous chapters is significantly broader. First, Craver focuses on the role of black boxes in covering ignorance. In contrast, I stressed that there are other, equally important reasons for introducing a black box, from drawing methodological boundaries to isolating a field from "external" critique. Second, and more important, my conception of black box *qua* placeholder in a causal explanation represented in a model also encompasses Hanson's "gray" and "glass" boxes. Allow me to elaborate.

---

[11] The notion of mechanism sketches is developed in Piccinini and Craver (2011).

Begin by comparing the black-boxing strategy with the "more detail is always better" approach. From this standpoint, adding *any* kind of specificity to a representation invariably increases its explanatory power. The outcome of this process of including micro-constituents, taken to the limit, is a truly complete mechanistic model that contains no black boxes or gray boxes. All complex components have been broken down to their fundamental building blocks, whatever that may mean. Thus understood, the mechanistic approach is fundamentally at odds with the black-boxing strategy advanced in this book. It roughly corresponds to my crude characterization of radical reductionism, from the first two chapters.

Nevertheless, it should be stressed that this "the more detail the better" approach is *not* the perspective articulated and defended by Craver and his collaborators. Indeed, Craver and Kaplan (2020) explicitly reject the "all details are necessary" thesis as "ridiculously stringent" (p. 310). They clarify that their position should rather be understood along the lines of a "more *relevant* details are better," according to which "if model M contains more explanatory relevant details than M^* about the [Salmon-Completeness] mechanism for P versus P', then M has more explanatory force than M^* for P versus P', all things equal" (2020, p. 303).

We need not delve into the specifics of the "3M" approach.[12] The important point, for present purposes, is that, if not all details matter, then there will be some aspects of the mechanism that are inevitably left out of the representation. Determining which features are relevant and which are not requires the three phases characterized in detail in Chapter 5—framing, difference-making, and representation—or something along those lines. What this goes to show is that my black-boxing recipe does not merely encompass what is commonly described as a "black box," that is, a representation where *no* underlying feature has been uncovered. In addition, it also captures placeholders where the underlying mechanisms or pattern of behavior is only partially understood, or whose representation is more or less deliberately left incomplete. These correspond to the Hanson-Craver gray

---

[12] Simply put, "3M" captures the "model-to-mechanism mapping" requirement that makes the commitment to a mechanistic view of explanation explicit: "In successful explanatory models in cognitive and systems neuroscience (a) the variables in the model correspond to components, activities, properties, and organizational features of the target mechanism that produces, maintains, or underlies the phenomenon, and (b) the (perhaps mathematical) dependencies posited among these variables in the model correspond to the (perhaps quantifiable) causal relations among the components of the target mechanism" (Kaplan and Craver 2011, p. 611). For a more nuanced, contrastive formulation, see Craver and Kaplan (2020, p. 297).

boxes and, indeed, their "glass boxes," once these are rightly understood in a contextual fashion.

Another influential discussion of black boxes has been provided by Strevens in his book *Depth*, where he articulates the "kairetic" strategy, the account of causal explanation outlined in Chapter 5. Strevens claims that explanations containing black boxes do not stand on their own. To make his point, he begins by distinguishing two ways in which a scientific explanation can be treated as "standalone." First, an explanation stands alone if it specifies all the difference-makers for its explanandum. An explanation containing black boxes may technically be complete in this former sense, as long as the black boxes stand in for the mechanisms that produce the explanandum event. A second, stronger kind of standalone explanation provides a full understanding of the explanandum in an unqualified, absolute fashion. A "deep" standalone explanation of this kind provides an exhaustive understanding of the explanandum.[13] Deep standalone explanations contain no black boxes. All filler terms have been unraveled; all mysteries unveiled. Thus, on Strevens's picture, explanations containing black boxes are technically complete, but only in the former, derivative sense. Black-boxed explanations, at best, confer partial, qualified, understanding of their explanandum. This is because they require the support of a mechanism-containing framework in order to stand alone. These technical remarks call for some clarification. What exactly does he have in mind?

Strevens clarifies the sense in which black boxes are only complete in a derivative sense by distinguishing two situations, depending on whether or not the black box in question is embedded in an explanatory framework. Focus, initially, on the latter scenario, where the black box is not framed. Strevens argues that a causal model containing a black box outside of the context of a framework is deficient, for two reasons. First, since—on Strevens's view, not mine—black boxes are causally empty, they have no causal powers. Black boxes cannot entail their target because they lack the capacity to produce anything; they have no mechanistic components. Second, since black boxes are multiply realizable, they are typically incohesive. The range of systems that black boxes stand in for is so broad that the box itself does not constitute a homogeneous entity. For these reasons, an explanation containing a

---

[13] The limiting case of these "deep" standalone explanations is what Strevens calls "exhaustive" standalone explanations, that is, maximally elongated, possibly infinite, indefinitely large characterizations of the causal history of the explanandum. There seem to be some analogies here with Craver's notion of a "complete mechanistic model."

black box does not stand alone. Only a "deep standalone explanation," which contains no black boxes and exhaustively describes all of its fundamental constituents, can survive outside of a framework.

Now, shift to the second, alternative scenario. When a black box is understood as a placeholder for a mechanism within the context of a framework, the two problems discussed in the previous paragraph do not subsist. This is because the box stands in for a specific mechanism, specified by the framework. This makes the black box cohesive, and it bestows upon it some causal capacities. Thus, Strevens claims, assuming that everything else is in order, an explanation containing a black box may stand alone. Thanks to the tacit support of the framework, and in virtue of the causal powers of the underlying homogeneous mechanism, black-boxed explanations can be causal, cohesive, and complete. There is, however, a catch. In these circumstances, what is being explained is not an explanandum *simpliciter*, an absolute event. Rather, the object of explanation is the occurrence of something *given that* a certain mechanism is in place. The framework-relativity of the explanation thus limits the explanatory power of the model. To explain a phenomenon, given that so-and-so is in place, Strevens claims, is a lesser achievement than explaining a phenomenon in an absolute sense. The ultimate goal of science is unqualified understanding, which can only be achieved via framework-independent explanatory models. Once the interior of all black boxes is illuminated, their darkness will finally be dispelled.

Time for a brief summary. Strevens's and Craver's positions on scientific explanation diverge in several important respects. For instance, while Strevens advances an overtly reductionist stance, Craver stresses the antireductionist flavor of his mechanistic approach as an integrative perspective. Correspondingly, their analyses of black boxes go in different directions. Nevertheless, they share some important features. For one, both philosophers treat black boxes as placeholders standing in for mechanisms. Second, on both views, black boxes seem to have no "productive" role to play in scientific practice. Sure, when used correctly, they allow researchers to proceed in the face of ignorance, postponing further inquiry until more auspicious times. When things go south, they become an impediment to true comprehension. Either way, we are always better off when a black box is replaced by a gray box and, eventually, with a glass one. In other words, on both accounts, black-boxing is something that we must come to terms with. But it is unclear that willingly introducing a black box brings any added value to scientific knowledge and understanding. In Chapter 10, I shall argue that black boxes play a productive role in science. In the meantime, let me stress three differences that set my analysis aside from these illustrious predecessors. As I will try to argue, these disanalogies stem from my characterization of mechanisms as epistemic throughout this chapter.

First, let me address a relatively minor disagreement. Strevens claims that "to black-box an unknown mechanism is to offer an explanatory hostage to fortune" (2008, p. 154). The reason is that, according to Strevens, an explanatory model containing a black box is successful only if there turns out to be a single, cohesive, causal mechanism standing behind the black box itself. [14] In contrast, on the view defended here, the success of a black-boxed explanation does not depend on the presence of a single mechanism. Sure, cohesiveness is always a welcome virtue, which makes explanations more elegantly unified. But the explanatory success of a black box does not hinge on it. This is because the number and kind of mechanisms depends inherently on the level of description. In addition, Strevens claims that a black box without a framework is causally empty and incohesive. I take it a step further: a black box *sans* framework isn't a black box at all!

Second, Strevens maintains, a placeholder, in and of itself, has no causal relevance. This, he argues, resolves the long-standing puzzle of how it is possible for a property, defined partly in terms of its ability to cause a particular event, to nevertheless usefully appear in a causal explanation of that same event. The solution suggested by Strevens is that the property in question does not partake in any causal explanation by means of its own dubious causal powers. Rather, it does so in virtue of being a surrogate for something with real causal "oomph." This position can be reconciled with my own account. One could certainly claim that fitness has no causal power of its own. As such, fitness only drives evolutionary forces in virtue of the underlying properties it instantiates. At the same time, I should also stress that my view does *not* presuppose or invite any causal fundamentalism of this sort, according to which true causal relations only occur at the lowest physical level and all that is left at higher levels is a sort of explanatory relevance. In blackboxing a mechanism, we bestow upon it a genuine causal role, *relative to a particular model*. To be sure, compared to a different level of description, the causal "oomph" may lie elsewhere. But, from an evolutionary perspective, fitness can be treated as having bona fide causal powers.

---

[14] To be precise, Strevens notes, this becomes more of a problem when one is explaining a regularity or a generalization, as opposed to an individual event, where the presence of a cohesive causal mechanism can be usually taken for granted.

This brings us to the third and most significant difference. Strevens and Craver both maintain that explanations containing black boxes are not exhaustive. But exhaustive with respect to what? Presumably, relative to the underlying layer of mechanisms that, if described in detail, would render the explanation complete. It should now be evident why this idea makes little sense from my perspective. If the difference between mechanisms and black boxes is only a difference in levels of description, replacing the former with the latter will make the explanation no more exhaustive or complete. If all scientific explanations are perspectival, they will inevitably contain boxes. And the presence of a black box in a model does not necessarily signal the incompleteness of the model in question. Now, to be sure, we can always open the box and provide more detail. Yet, as we shall see in Chapter 10, this presupposes a shift in explanandum. We are not invariably explaining more and better. We are explaining something altogether different.

One final remark. What about mechanistic philosophers, who explicitly advocate a less-ontological mechanistic approach? Is my "diet" mechanistic philosophy just a version of the metaphysically sober, epistemic conception of mechanisms advanced by Bechtel and his collaborators? It seems to me that the answer is negative. While traditional epistemic mechanism explicitly dampens its ontological implications, it remains committed to controversial mechanistic tenets, such as decomposition and localization.[15] In contrast, the diet approach, which treats mechanisms as placeholders, also relaxes these methodological constraints. While my black-boxing recipe recognizes mechanistic explanation as one form of explanation in science, it poses hardly any strictures on how it works. That is what ultimately distinguishes it from extant forms of mechanistic philosophy.

In conclusion, we can all agree that black boxes are placeholders. These placeholders may stand in for mechanisms (difference-makers), but they can also stand in place of patterns of behavior in need of explanation (frames). They may be introduced because of ignorance, but also to implement and promote a specific research program. As I have defined them—placeholders in causal explanations represented in models—black boxes are broad, encompassing what are traditionally conceived as black, gray, and glass boxes. Black boxes are mechanisms, and may also stand in for mechanisms, as long as we relativize black boxes and mechanisms to different explanatory levels. In this respect, I disagree with both Strevens's idea of a "standalone"

---

[15] For a discussion of these tenets, see Silberstein and Chemero (2013).

explanation and Craver's "ontic" conception of explanation. The key is to recognize that all scientific inquiry is framed and qualified. Unqualified understanding and completeness are never to be found in actual scientific practice. At best, they are regulative ideals.

In the words of a prominent twentieth-century statistician, "Since all models are wrong the scientist cannot obtain a 'correct' one by excessive elaboration. On the contrary following William of Occam he should seek an economical description of natural phenomena. Just as the ability to devise simple but evocative models is the signature of the great scientist so overelaboration and overparameterization is often the mark of mediocrity" (Box 1976, p. 72)—*nomen omen!* We shall explore the implications of this claim in Chapter 10. Before doing so, our next task is to wrap up our discussion of mechanisms by exploring some implications of my proposal.

§7.6. What's in a Mechanism?

Section 7.4 argued that the concept of mechanism is better understood as a model-theoretic construct, not to be confused or conflated with physical entities and processes interacting in the universe. What are the advantages of this approach? This section shows that my "diet" mechanistic philosophy successfully addresses issues with traditional mechanistic approaches.

One complaint about the new wave of mechanistic philosophy is that its key terms—entity, activity, complex system, and the like—are slippery, rendering the definiens, the concept of mechanism, hopelessly vague and ambiguous.[16] After all, virtually anything can be characterized as "organized aggregation of entities and activities collectively producing a specific phenomenon." A concept that applies to almost everything seems trivial.

If, as some philosophers suggest, *mechanism* is supposed to denote a true ontological category—some would say, a natural kind—then the worry becomes serious. Unless one provides a clear-cut definition of what exactly counts as a "mechanism," its extension becomes a hodgepodge of systems, entities, and processes with little in common except, perhaps, some generic claim that they make something happen. What use is such a wildly heterogeneous ontological category for scientific and other empirical practices?

---

[16] For a general overview and discussion of this difficulty, see Boniolo (2013).

From our diet perspective, mechanisms do not constitute a natural kind at all. They are a pragmatic classification. We call "mechanism" the model-theoretic representation of entities or processes that produce an effect that we are interested in studying. This preliminary description is a placeholder that represents a set of behaviors or the activities that bring them about. The structure of non-fundamental mechanisms, the mechanisms described in the special sciences, can always be further unpacked, revealing its status as placeholders, as black boxes. In short, mechanisms play a very important role within theory and practice. But *not* as an ontological category.

This brings us to a second concern. Nicholson (2012) provides a historical overview concluding that, over time, the term "mechanism" has acquired various meanings and uses in biology. Specifically, Nicholson argues that "mechanism" is used in the biological sciences in three senses. First, it may refer to what he calls *mechanicism*, the philosophical conception of living organisms as machines that can be exhaustively explained in terms of the structure and interaction of their component parts. Second, "mechanism" may refer to *machine mechanism*, that is, the collective internal workings of a machine-like structure. Third, and finally, "mechanism" sometimes refers to *causal mechanism*: a step-by-step explanation of the mode of operation of a causal process that gives rise to a phenomenon of interest.

The heterogeneity of "mechanism"—a concept with a rich and nuanced history—seems both plausible and unsurprising. But Nicholson's point, as I see it, is not merely the vagueness and ambiguity acknowledged in the previous paragraphs. His critique and ensuing suggestions are based on a consequence of semantic conflation. The problem, in short, is that mechanisms often end up wrongly classified as an ontological category. This has unpalatable consequences. As he puts it, "Mechanismic philosophers tend to conceive mechanisms as real things in the world existing independently from our conceptualization of them" (2012, p. 158). In contrast, he argues, "causal mechanisms are better understood as heuristic explanatory devices than as real things in nature, and [...] the reason why most mechanistic philosophers think otherwise is because they inadvertently transpose the ontic status of machine mechanisms onto their analyses of causal mechanisms" (p. 152).

I concur with both the diagnosis of the problem and the follow-up. As noted in the previous pages, it is unfortunate that the same term, "mechanism," is used to refer to both stuff in the world and to the model-theoretic representation of this stuff. The standard distinction between mechanisms and mechanistic models goes some way toward alleviating the confusion.

But it hardly goes far enough. Furthermore, I agree with Nicholson that the scientifically important notion is the concept of *causal mechanism*, which includes heuristic devices that facilitate the explanation of phenomena. The abstract and idealized entities postulated in models are often quite different from the entities that are represented, and the two must be kept separate.

My solution is simple. Draw an explicit distinction between real entities and processes in the world, on the one hand, and mechanisms, understood as theoretical representations, on the other. By treating mechanisms as placeholders in causal explanations framed in models, we can eschew the confusion noted by Nicholson. Characterized as placeholders, mechanisms shed their ontological implications. They become boxes: black, gray, or glass, depending on the amount of detail represented or idealized away.

Finally, let us address a third concern. In a recent article, Franklin-Hall (2016) provides a general assessment of the philosophical impact of the mechanistic wave. Judging by both the language of the new mechanists and the influence of their work, one would legitimately expect that the new mechanism has served up a bevy of solutions. In contrast, she argues, with respect to the central task of elucidating the nature of explanation, the movement as a whole has hardly delivered on its promises. Interestingly, her critique is not centered on the falsity or incorrectness of the central mechanistic tenets. Rather, the main concern is that "mechanistic explanatory accounts offered to date—even in their strongest formulations—have failed to move beyond the simple and uncontroversial slogan 'some explanations show how things work.' In particular, I argue that proposed constraints on mechanistic explanation are not up to the task required of them: namely, that of distinguishing acceptable explanations from those that, though minimally mechanistic, are incontrovertibly inadequate" (2016, p. 42).

Franklin-Hall concludes that, although the new mechanistic attempt to take scientific practice seriously—and, more generally, its naturalistic stance—is important and admirable, much remains to be done. We need to keep looking under the hood of explanatory practice and detail its workings. The reason is that "rather than opening the black boxes of the scientific enterprise—with respect to causation, part individuation, and explanatory level—philosophers have largely taken those practices for granted" (2016, p. 71).

I wholeheartedly agree. Describing the mechanistic structure of a system is only one step in a long and tortuous stride. Prediction, explanation, abduction, and other inferences require framing, identifying difference-makers, abstraction, idealization, modeling, and many other forms of representation.

The focus should be on the nature of these epistemic practices, not on the notion of mechanism, characterized as an ontological category.

There is one point of detail where I depart from Franklin-Hall's assessment. In her final footnote, she writes: "There is one mildly ironic exception to my general diagnosis. The only putative black box that mechanists have opened is the scientists' concept of 'mechanism.' On reflection, this focus was imprudent. Not every concept used by scientists is meaty, and not every term reflects a genuine black box; 'mechanism' is not a theoretical term within the science, but a mere pointer, a placeholder—similar perhaps to the philosopher's term 'conception'" (2016, p. 71, fn. 17). I would rephrase the point in slightly different terms. A mechanism is, indeed, a pointer, or placeholder. But, precisely for this reason, it is also a black box: a placeholder that stands in for a range of behaviors or whatever produces it. Once again, the power of black boxes should not be underestimated.

§7.7. The Diet Recipe

In conclusion, it is time to tie up some loose ends. The new mechanistic philosophy is a welcome contribution to the philosophy of science. First, it has promoted a healthy naturalistic attitude, which calls philosophers to draw attention to the details of actual scientific practice. In addition, it offers a further, lucid analysis of why physics should not be heralded as *the* paradigm of science. Mechanistic disciplines, such as biology and neuropsychology, are inherently different from mathematized and law-based physics. My black-boxing approach owes a substantial intellectual debt to the new mechanists. Still, framing, difference-making, and representation draw a wedge between ontology and epistemology, which do not go hand in hand, as it is too frequently assumed. Hence, we should not use the same term—"mechanism"— to refer to both. In short, black-boxing tempers the ontological implications of mechanistic approaches to science.

At the root of the trouble lies the notion of mechanism itself. Mechanisms have a two-sided nature. On the one hand, philosophers tend to adopt an ontology according to which mechanisms are a fundamental category or natural kind: they are real stuff in the real world. On the other hand, mechanisms constitute a model-theoretic construction; they are a vehicle of scientific representation. From this latter standpoint, neo-mechanists emphasize the significance of discovering, modeling, and studying these entities. Mechanistic ontology and mechanistic epistemology are commonly treated as two sides of the same coin. This seems intuitive enough. If mechanisms are the cement of the universe, then it is hardly surprising that science seeks to discover and represent their structure.

In contrast, I argued that a single, well-defined notion of mechanism cannot play both roles at once. Metaphysical and epistemic considerations require distinct conceptions of mechanism. The general problem is that explanations and descriptions do not always proceed hand in hand—a point famously raised by Cartwright, but often overlooked. Explanation is a paramount scientific goal. So is description. But one cannot generally fulfill both endeavors at once. The black-boxing strategy is an attempt to capture why this is the case. Truth and explanation typically presuppose different abstractions and idealizations in the representation of frames and difference-makers. A single model, no matter how general or overarching, is not going to cut it.

My proposal involves a radical separation of the two dimensions of mechanisms—metaphysics and epistemology. Mechanisms, suitably represented and embedded in models, play a central role in the construction of black boxes and, more generally, within scientific practice. But they do not constitute a natural kind. For all we know, there may well be an ontological category of "mechanism," or something akin to it, in the world. Still, even assuming that such a homogeneous group exists, it does not correspond to the model-theoretic construct at the heart of scientific practice. Mechanisms are placeholders. They stand in for entities that produce a phenomenon. This role is epistemic. For all these reasons, I playfully refer to black-boxing as a form of "diet" mechanistic philosophy, with all the epistemic flavor of the traditional views, but no ontological calories.

8

Emergence Reframed

> In general, nature does not prepare situations to fit the kinds of
> mathematical theories we hanker for. We construct both the theo-
> ries and the objects to which they apply, then match them piecemeal
> onto real situations, deriving—sometimes with great precision—a
> bit of what happens, but generally not getting all the facts straight
> at once.
>
> —Nancy Cartwright, *How the Laws of Physics Lie*, p. 162

§8.1. Introduction

Let us retrace our steps back to the early stages of our journey. Reductionism
contends that science invariably advances by descending to lower
levels. Antireductionism flatly rejects this tenet. Some explanations,
antireductionists counter, cannot be enhanced by breaking them down any
further. But why should this be so? What makes explanations "autonomous"?
A popular way of cashing out the autonomy thesis, the core of antireductionism, involves the concept of *emergence*.

The main intuition underlying emergence is simple. As systems become
increasingly complex, they begin to display properties which, *in some sense*,
transcend the properties of their parts. As such, they exhibit behavior that
cannot be predicted by, explained with, or reduced to laws that govern simpler entities. Common wisdom shows that there is more to a team than just
a bunch of players, and a musical symphony is not merely an arrangement
of notes. The main task of a philosophical analysis of emergence is to spell
out the "in some sense" qualifier. In what ways, if any, do emergents transcend aggregative properties of their constituents? How should one understand the alleged unpredictability, non-explainability, or irreducibility of the
resulting behaviors of teamwork, musical compositions, and other sophisticated ensembles? Answering these questions might look simple. But it has
challenged scientists and philosophers alike for quite some time.

Black Boxes. Marco J. Nathan, Oxford University Press. © Oxford University Press 2021.
DOI: 10.1093/oso/9780190095482.003.0008

The general idea of wholes transcending the sums of their parts is hardly novel. It can already be found in Aristotle's *Metaphysics*. More modern discussions of emergence trace their roots back to J. S. Mill's distinction between "homopathic" and "heteropathic" laws and have been shaped into their contemporary form in the early 1900s by the work of philosophers such as Broad, Alexander, and Lovejoy, and scientists like Morgan and Sperry. Set aside during the heyday of logical positivism, when it was regularly dismissed as confused and teetering on incoherence, over the last few decades, emergence has worked its way back into the mainstream.

The comeback of emergence onto the main stage has had a polarizing effect, dividing the camp into enthusiastic supporters and skeptical naysayers. Authors with antireductionist tendencies, scientists and philosophers, appeal to emergence to capture a form of non-reductive physicalism. Reductionists, on the other hand, treat emergence as an obscure, muddled notion that should be definitively confined to oblivion. Once again, we are stuck between Scylla and Charybdis. But is this the only path? Is emergence truly incompatible with reduction? I suggest a negative answer.

This chapter presents, motivates, and defends a recharacterization of emergence and its role in scientific research, grounded in our analysis of black boxes. Here is the plan. Section 8.2 sets the stage by providing an overview of extant accounts. Section 8.3 examines some ways in which emergence is employed in complexity science and concludes that current definitions have trouble accommodating such uses. Readers previously familiar with the debate should consider skipping directly to section 8.4, which offers a constructive proposal. Emergents, I maintain, are best conceived as black boxes: placeholders in causal explanations represented in models. My suggestion has the welcome implications of bringing together various usages of emergence across domains and reconciling emergence with reduction. This, however, does come at a cost. It requires abandoning a rigid perspective according to which emergence is an intrinsic or absolute feature of systems, in favor of a more contextual approach that relativizes the emergent status of a property or behavior to a specific explanatory frame of reference. The final sections discuss refinements and implications (§8.5), as well as a few adequacy conditions and advantages (§8.6) of the proposal.

Before moving on, two disclaimers are in order. First, my proposal does not presuppose that all accounts of emergence should be judged from a scientific perspective or, more generally, that science provides the ultimate standards for philosophy. I simply maintain that making sense of scientific concepts is a worthwhile philosophical endeavor, and emergence is no exception. Second, my suggestion is not to discard and replace all existing definitions of emergence. My contention is that many extant analyses have locked themselves into conceptual straitjackets by unnecessarily presupposing that emergence is either an absolute feature of systems or is completely dependent on the current state of our knowledge. By dropping these assumptions and rejecting the emergence vs. reduction contrast, we make room for a definition that reconciles seemingly incompatible aspects of emergence, providing a more flexible, unified, and ecumenical framework.

§8.2. Emergence: Metaphysical or Epistemic?

To cash out the general thesis underlying emergence, it is helpful to return to the rough, context-dependent partition of the natural world into distinct levels, first introduced in section 1.2 of Chapter 1. Our "wedding cake model," illustrated in Figure 1.1, depicted physics, biology, neuropsychology, and economics, arranged in terms of constitution, with smaller constituents at the bottom and larger-scale entities arranged toward the top. With this image in mind, emergents can be characterized as higher-level properties that, in some sense, are both *dependent* and *autonomous* from their underlying lower-level basis. Intuitive as it may seem, this preliminary definition is inadequate, for two reasons. First, the combination of dependence and autonomy generates an obvious tension: are these notions not contradictory? Second, as noted at the outset (§8.1), the "in some sense" qualifier renders the analysis hopelessly vague, bordering on vacuity. An account of emergence worth its salt must clarify our first-pass analysis on both scores.

Our "wedding cake" layering of reality into distinct levels can be interpreted either metaphysically or epistemically. This suggests two general strategies for refining our preliminary sketch. On the one hand, "metaphysical" approaches maintain that emergents are not exhausted, ontologically, by their constituents. Wholes are temporally dependent (synchronically or diachronically[1]) and causally, modally, or nomologically autonomous from parts. "Epistemic" readings, on the other hand, typically agree with their metaphysical counterparts on the temporal dependence of emergents, but

---

[1] While I focus on synchronic dependence between emergents and their basis, these relations can also be characterized diachronically or dynamically (Humphreys 2016).

differ in treating wholes as epistemically, as opposed to ontologically, autonomous from parts. This section outlines both families of views, in turn.

A popular strategy for cashing out metaphysical emergence, which goes
all the way back to British emergentism of the early 1900s, is to analyze
"ontological autonomy" in terms of non-reducibility.[2] On this view, unlike properties that are merely resultant, emergents are not explainable or
predictable from the lower-level constituents from which they emerge, so-
called basal conditions. This is because emergent properties are, by definition, *irreducible*.

Compelling as it may sound, this "supervenience emergentism" faces
powerful objections.[3] In work spanning a few decades, Jaegwon Kim
has argued that irreducibility runs the risk of trivializing emergence
by turning it into an empty doctrine. According to Kim, all properties are either physical properties or reducible to physical properties via
functionalization.[4] From this perspective, there seems to be no room for
genuine emergents, with the possible exception of subjective states of consciousness ("qualia") and other properties that fall outside the scope of
current scientific inquiry.

In short, the main question underlying supervenience emergentism is:
can emergent properties be reduced to their bases? Emergentists argue
in the negative. Critics, such as Kim, respond that such reduction is feasible, at least in principle. Kim's influential argument has raised a host of
reactions. Is specifying the physical realization of a functional description sufficient for reduction? Is there a unique and complete lower-level
characterization of all higher-level phenomena? These questions should
have a familiar ring to them, as they mirror the stalemate between friends
and foes of reductionism depicted in section 2.6 of Chapter 2. And the
prospects for a positive resolution seem no better here. The concept of reduction, alas, is no clearer than emergence. Consequently, purporting to
explicate the latter in terms of the former will get us out of the frying pan
and right back into the fire.

One solution is to drop the problematic appeal to reduction and to define emergence in terms of other—allegedly less slippery—ontological

---

[2] An excellent overview of early emergentism can be found in McLaughlin (1992).

[3] The view under present scrutiny is sometimes dubbed "supervenience emergentism" because it
depicts emergence as a brute supervenience relation holding in virtue of fundamental emergent laws.

[4] As we saw in section 2.6, "functionalizing" a property H involves describing H in terms of inputs
and outputs, and then showing how H's causal-role description is realized at a lower level. This influential strategy is developed in detail in Kim (1999, 2005).

concepts, such as *causal powers*. One may thus distinguish two forms of metaphysical emergence. "Strong emergence" occurs when higher-level properties synchronically depend upon their basal conditions and, yet, these macro-properties display distinctive causal powers that are absent from lower levels. A less demanding conception, which fits in better with the physicalist outlook by eschewing controversial forms of "downward causation," retains the synchronic clause. Yet, it replaces the assumption that emergents have novel causal powers with the requirement that the powers of emergents must be a proper subset of the set of powers found at the basal level. "Weak emergence," thus defined, avoids the controversial implication that new causes may arise at higher levels. Still, it purports to vindicate a form of non-reductive physicalism, as higher and lower levels can be distinguished via an application of the indiscernibility of identicals principle.[5]

Intuitive as they may seem, definitions of emergence appealing to causal powers are hardly unproblematic. For one, the precise individuation of powers constitutes a thorny challenge. The issue is that virtually any description of any phenomenon involves properties, causal or otherwise, that are lost by ascending or descending levels. How do we determine "novelty," and on what basis? Are the liquidity of water and the elasticity of rubber "distinctive" powers, or do they restate the capacities of their constituent molecules? The bottom line is that, in spite of the struggle to avoid the vagueness of reductionism, appeals to causal powers are plagued by at-root identical problems: properties and their linguistic descriptions do not wear their level and status on their sleeve. Consequently, providing a set of clear-cut individuation criteria and illuminating analyses is no trivial endeavor.

We shall return to causal powers and metaphysical strategies in sections 8.3 and 8.4. Before doing so, let us introduce the alternative route.

The most influential epistemic analysis of emergence was offered by Hempel and Oppenheim, who reject the idea of emergence as an ontological property inherent to specific systems. They characterize emergence as an epistemological stance, indicative of the scope of our knowledge at a given

---

[5] Simply put, the *indiscernibility of identicals* states that identical entities must share all properties. Thus, if some object *a* lacks some property of object *b*, the two objects *a* and *b* can be deemed ontologically distinct.

time, thus making it extrinsic or relational, as opposed to intrinsic or absolute in character.

> The occurrence of a characteristic W in an object w is emergent relative to a theory T, a part relation [i.e., a part-hood relation] P_t, and a class G of attributes if that occurrence cannot be deduced by means of T from a characterization of the P_t-parts of w with respect to all attributes in G. (Hempel and Oppenheim 1965, p. 64)

Epistemic accounts of emergence remain popular. Bedau (1997, 2002) spells out a form of weak emergence characterized by "non-derivability *sans* simulation." An effective illustration is Conway's game of life, where the behavior of the higher-level patterns is completely determined by the micro-state and yet, because of computational complexity, one cannot predict how the system will evolve over time without running a computer simulation.

Others refine the original insight by rendering emergence sensitive to the kind of explanation offered. Hempel and Oppenheim treat emergence as lack of explanation *tout court*. Yet, one may allow the same event to be emergent with respect to one kind of explanation, say, mechanistic explanation, but not with respect to a different kind, say, functional explanation.[6]

Time to take stock. Modern accounts of emergence come in two families: metaphysical and epistemic.[7] Traditional metaphysical approaches purport to cash out ontological autonomy in terms of reducibility. More recent endeavors replace the focus on reducibility with causal powers, fundamentality, or necessitation.[8] Epistemic accounts define emergence in terms of how complex systems may (not) be explained. Differences should not be exaggerated. Many metaphysical accounts take epistemic features as indicative of emergence. Conversely, explanation, or lack thereof, may often depend on the nature of the system. Still, metaphysical and epistemic approaches disagree on the constitutive features of emergence and, hence, they are best kept distinct.

Our overview of philosophical discussions of emergence barely scratches the surfaces of a rich, long-standing, and nuanced debate. My sketchy outline

---

[6] For a proposal along these lines, see Taylor (2015).

[7] Humphreys (2016) prefers a trichotomy between *ontological*, *inferential*, and *conceptual* accounts of emergence. An even finer-grained categorization can be found in Gillett (2016). I opt for a slightly simpler taxonomy.

[8] See, respectively, Wilson (2021); Barnes (2012); and Chalmers (2006).

admittedly fails to do justice to the extant and growing literature on the topic. Still, I hope to have done enough to motivate the significance of two general methodological questions. First, do extant accounts capture the multifarious ways in which emergence is employed across disciplines, effectively providing a unifying definition? Second, if they do not, should we keep using a laden concept like "emergence," or would we be better off replacing it with allegedly clearer and less controversial expressions, such as "unpredictable" and "unexplainable"? The next two sections address both issues, in turn.

§8.3. Emergence in Action: Systems Neuroscience

The previous section suggested that providing a clear, compelling, and uncontroversial definition of emergence is no trivial endeavor. There are several promising accounts currently on the market, which I have categorized as metaphysical vs. epistemic. While all of them pinpoint important aspects of emergence, the question is whether any one is general and well-rounded enough to fully capture this historically elusive concept.

Despite the lack of an uncontroversial, widely accepted analysis, the term "emergence" continues to be employed extensively both in philosophy and across the natural and social sciences. Fascinating examples come from complexity theory, a growing discipline which encompasses a diverse array of interdisciplinary approaches aimed at understanding and explaining sophisticated biological and psychological systems. This section introduces and discusses some illustrations. Our goal is to examine whether, and to what extent, the theoretical accounts of emergence surveyed in the preceding section map onto empirical practice.

To get the ball rolling, let us focus our attention on a specific subfield of complexity science: systems neuroscience. In a contribution to a recent volume discussing prospects and challenges of contemporary neuroscience, Olaf Sporns (2015) argues that, to fully understand the workings of the brain, neuroscience must "shift perspective, towards embracing a view that squarely acknowledges the brain as a complex networked system, with many levels of organization, from cells to cognition that are individually irreducible and mutually interconnected" (p. 90). What motivates this shift in perspective, especially given the historically successful record of brain studies? Sporns acknowledges that the so-called neuron doctrine—the hallowed tenet that neurons are the fundamental computational units in the brain—has provided much insight. Yet, it is now evident that the power of neurons derives from their collective action as parts of networks, bound together by connections which facilitate their interaction, competition, and cooperation. Comprehending the brain requires grasping these interconnections.

Sporns describes this breakthrough, his envisioned shift in perspective, as the replacement of an old-fashioned account of neural circuits with a more sophisticated alternative. Traditionally, the actions of a neural circuit were assumed to be fully determined by the additive sum of highly specific point-to-point exchanges of information among its elements, typically, individual neurons. Consequently, the structure of the entire circuit was taken to be decomposable—in principle, given complete data—into neat sequences of individual causes and effects. Paraphrasing this into philosophical jargon, the overall function of a circuit was assumed to be the resultant of the collective interactions of individual components. In contrast, Sporns maintains, cutting-edge approaches to complexity theory and network neuroscience emphasize how global outcomes cannot be broken down into aggregates of localized causes and how the functioning of the system as a whole transcends the functioning of each individual element. In short, the lesson that neuroscience imported from systems biology is that the behavior of cells depends on gene regulatory networks, signal networks, and metabolic pathways, which shape and govern interactions among individual molecules.

How is any of this relevant to our discussion? The answer is that, at this point in Sporns's presentation, our old friend emergence enters the picture. The proposed shift in perspective concerning neural circuits purports to capture configurations of complex networks, which do not arise at lower levels of organization. These are typically global states of brain dynamics in which huge collections of neurons engage in coherent and collective behavior through local interactions, individually quite weak, but aggregatively powerful enough to generate large-scale patterns. Sporns mentions "neural synchronization"—the coordinated, synchronized firing of large numbers of neurons—and the "critical state," a dynamic regime where systems engage in a wide range of flexible and variable tasks. These, he claims, are *emergent* phenomena.

Unsurprisingly, these neural processes become quite complex very fast. Technical details need not concern us here.[9] The question that I want to address is a philosophical one. Can we adequately conceptualize the emergence

---

[9] For an extensive but readable overview of these findings, see Sporns (2011, 2012).

of phenomena, such as neural synchronization and the so-called critical state, in terms of irreducibility, causal powers, unpredictability, or lack of explanation, that is, along the lines drawn by extant accounts of emergence?

Prima facie, there seem to be reasons for cautious optimism. Statements along the lines of Sporns's remarks that the functioning of complex neural networks as a whole "transcends the functioning of each of its individual elements" (2015, p. 92) and "give rise to new properties that do not exist or arise at new levels of organization" (2015, p. 93) seem to fit in well with both metaphysical and epistemic accounts of emergence. Yet, upon further scrutiny, complications arise, and the picture becomes less promising.

Focus, first, on classical epistemic analyses, such as Hempel and Oppenheim's, which characterize emergence as an overarching lack of explanation. Despite consistent progress, many aspects of the human brain remain deeply hidden mysteries. At the same time, synchronization, critical states, and other modes of connectivity do not seem to emerge from lack of explanation *tout court*. After all, these systems have been extensively studied and, challenges notwithstanding, we do have detailed, if only partial, accounts. Indeed, the mapping of a complete "connectome"—a wiring diagram of the brain at single neuron resolution—may soon become a reality.[10]

To further complicate things, the classic epistemic approach does not generalize nearly enough. Even if neural systems did emerge in virtue of the current unavailability of adequate explanation, it would be simply preposterous to attribute the emergence of, say, liquidity, isomers, protein folding, and many other paradigmatic examples to lack of analysis. Elementary chemistry textbooks provide clear and exhaustive accounts of isomers and how hydrogen and oxygen combine to produce water. These structural properties explain the transparency and liquidity of water, its thirst-quenching capacity, and other types of higher-level behavior.[11] In short, classic epistemic accounts à la Hempel and Oppenheim fail to justify the emergent status of neural circuits and various other scientific posits.

---

[10] See, for example, Sporns (2012) and Zador (2015).

[11] To circumvent the difficulty, one could insists that what makes these processes emergent is the absence of *complete* explanation. This move runs the risk of trivializing the entire account. The reason is that we lack a complete explanation of virtually any scientifically interesting phenomenon. The notion of "complete explanation" is rather murky. Sure, neural systems and protein structure exceed the grasp of a comprehensive description. But could one not say the same about gene expression, RNA synthesis, and other cytological processes? Should one label "emergent" anything that cannot be depicted exhaustively? When exactly, if ever, are we entitled to dub an explanation "complete"?

What about more sophisticated epistemic routes? For instance, following Bedau, one could maintain that complex brain states are only derivable via simulation. This, on his definition, makes them weakly emergent. Or, perhaps, could the missing component be an appropriate explanation of these neural states—a truly mechanistic description? I am sympathetic to both proposals, which capture important aspects of neural complexity. Nevertheless, explanation, or lack thereof, falls short of constituting the entire story. Emergence, as it is standardly used in science, has a dimension that transcends our current state of knowledge. Allow me to elaborate.

At the most fundamental level, what all epistemic accounts of emergents have in common—indeed, what makes them "epistemic" in the first place— is the presupposition that, were our mental states, cognitive capacities, or general state of knowledge different, then some emergent properties would no longer be emergent. To wit, if we could predict the effects of neural synchronization at a glance, without the aid of simulation, then, by Bedau's own lights, this phenomenon would not count as emergent. Similarly, if we could provide the appropriate kind of reconstruction, then, on all explanation-based accounts, emergent properties would *ipso facto* cease to be so. I do not doubt that many emergent properties cannot currently be explained, and the required kind of understanding may well exceed our grasp. But is this the decisive factor that makes them emergent? This seems disputable. Note how, despite Sporns's emphasis on explaining and understanding the brain, nothing he says makes the emergence of neural systems dependent on the state of scientific knowledge. The emergence of complex patterns is a feature of the brain itself. We can already explain a good deal about these states, and further discoveries would not change that such complexities only arise at higher levels of organization, in a non-additive fashion. In this respect, the emergent systems described by Sporns fit better with traditional ontological analyses than with their epistemic counterparts.

With this in mind, consider the prospects of reconstructing the emergence of neural and other scientific systems in metaphysical terms. To begin, if "irreducibility" is understood as "unpredictability" or "inexplicability," then we immediately run into all the difficulties just encountered. Given that neural synchronization, critical states, and protein folding can already be predicted and explained rather effectively, *a fortiori*, they are also predictable and explainable. Hence, let us focus on different ways of making sense of new properties arising at higher levels of organization.

Causal powers are a promising candidate, which provides precious resources to analyze the logic and commitments behind scientific arguments. In what follows, I set aside problems specific to this or other proposals, none of which is uncontroversial. Rather, I focus on a broader, more general issue. All attempts to analyze emergence in metaphysical terms share one shortcoming. They leave out the other side of the story: epistemology.

To clarify, let us return to Sporns's new frontier of neuroscience. Our understanding of the brain, he claims, is moving away from additive views, toward more systematic approaches that reject the identification of complexity with localized interactions between constituent links. Do these higher levels exhibit genuine novel powers, the signature of strong emergence, or a proper subset of more fundamental capacities, as required by weak emergence? As noted in section 8.2, trying to pinpoint powers and causal descriptions drags us back into the quagmire of the reductionism vs. antireductionism rhetoric which, I argued in Chapter 2, ends up being more verbal than substantial.

The key insight is better conceptualized along different lines. Networks are rapidly becoming the basic units of organization in the study of the brain. The point is not that we cannot break these systems down to their constituents and interactions—of course we can. But doing so is futile: it has little to no explanatory value. This interpretation of emergence eschews a pure ontological reading, based entirely on what neural systems *are*, toward an epistemic focus on how complex neural systems ought to be conceptualized and *explained*. This is where black boxes enter the stage.

Time to take stock. Current analyses fail to fully capture the use of emergence in complexity science. Of course, the philosophical moral of our examples is quite general. As prophetically foreseen by Russell (1927), metaphysical stances have trouble explaining why the emergent status of a scientific entity often involves theoretical shifts in perspective, explanatory refinements, and other forms of epistemic progress. At the same time, the focus on lack of explanation *tout court* or of a specific kind, which characterizes epistemic views, is not the full story either. The emergent status of an entity or property cannot be boiled down to the state of knowledge of specific agents or communities. In short, what is required is a conception that combines the flexibility of epistemic accounts with the mind-independence of ontological approaches. The next section suggests how both perspectives can be molded into a general analysis, based on our black-boxing recipe.

§8.4. Emergents and Black Boxes

Section 8.3 maintained that no extant philosophical analysis, either epistemic or metaphysical in character, fully captures the role of emergence in science. What ought one conclude from this? And how should we move forward?

One response is to relax the monistic presupposition and give up altogether on the project of finding a univocal concept of emergence that spans across the board. Developing a single definition that fixes the meaning of emergence across, say, physics, biology, cognitive science, and economics and connects it to all its uses in metaphysics, epistemology, ethics, and other areas of philosophy may well be an overly ambitious endeavor. But what exactly is the alternative? A more flexible, pluralistic approach that rejects a monolithic definition of emergence in favor of local differences still raises the question: in virtue of what are these variations of *emergence*?

This section sketches a proposal for systematizing various uses of emergence. As stressed at the outset (§8.1), it is not my intention to discard all previous insights and replace them with a brand new analysis. The aim is rather to provide a broader conceptual framework, a definition schema that leaves room for different ways in which phenomena may be emergent, while capturing the core features that all the overlapping uses have in common.

Here is the punchline. Contrary to many metaphysical accounts, what makes a property, system, or process "emergent" is independent of ontological or causal novelty. Lack of explanation, a core tenet of epistemic approaches, is not the decisive factor either. At the most general level, the characteristic of an emergent is an extrinsic relation to its basal condition. Property P is emergent, relative to a model, theory, or hypothesis H, if P is used in a causal explanation, in the context of H, without specifying further its nature, structure, or lower-level implementation. This, as we shall see, has both metaphysical *and* epistemic components. An emergent, in this sense, is a black box: a placeholder in a causal explanation represented in a model. I now clarify the proposal by illustrating it with examples.

Consider the folding of protein. Is it an emergent process? Extant accounts deliver conflicting verdicts. Is the behavior of amino acids irreducible, unpredictable, or unexplainable? Is it causally novel? The answer depends on how the explanandum is framed, explained, and represented. What degree of precision is expected from the depictions? How much structure are we allowed to assume? Can we idealize or abstract away some of the detail? It should be evident that, unless these preliminary questions are addressed, it becomes meaningless to ask whether the folding of protein is "emergent" along these lines or any other. The problem has nothing to do with the folding of protein per se. The core issue is the vagueness of notions like reduction, causal power, and novelty. Thus, the question remains: is the behavior of protein emergent? And, if it is, what makes it so?

Let us explore an altogether different strategy. Set aside the phenomenon of protein folding, in and of itself, and focus instead on how protein behavior figures in biochemical explanations. The action of amino acids is invoked in various molecular-biological explanatory contexts involving, for instance, gene regulation and embryonic development. Regardless of whether protein folding is an explanans or an explanandum, these depictions typically omit the details of why and how protein assumes its tertiary or quaternary structure. Is it because the relevant information is unavailable? No. Contemporary biochemistry does not treat amino acids as irreducible or unexplainable. Is it because the underlying micro-processes make no difference to the macro-process? Again, the answer is negative. The composition of the solvent, the concentration of salts, pH, temperature, cofactors, and molecular chaperones, to name a few cytological factors, are all crucial to the final structure of the protein. This is the outcome of a complex, stable process which requires a number of background conditions.

Then why are all these features often omitted? The reason is that, fascinating as they are, once the folding of protein is appropriately contextualized, these structural details are not required for describing the relevant frames and difference-makers. Sure, all these "gory details" contribute decisively to the folding of protein. Yet, they are neither relevant nor useful at a higher level of description. Protein folding is emergent because it is a black box.

None of this hinges on the specifics of protein folding. Analogous considerations can be applied, *mutatis mutandis*, to isomers, liquidity, language processing, and several other prime examples of emergence. All these phenomena play a substantive predictive and explanatory role across various contexts. The psychological capacity to process natural language, for instance, is a complex disposition that most humans acquire spontaneously during a specific developmental window or, with much greater effort, later in life. What makes the mastery of language "emergent" is neither the lack of physiological, evolutionary, or neuropsychological explanations of its nature, nor its alleged irreducibility. The core factor is rather that the—frustratingly elusive—details of the process are irrelevant for linguistic capacities to figure in macro-explanations, from folk-psychological scenarios ("Sam could not

follow the movie because she does not speak German") to serious path-
ological conditions such as aphasia, autism, or dyslexia. In short, what all
emergents have in common is that their micro-structure can be idealized and
abstracted away in the context of macro-explanations. They are black boxes
and this, I contend, makes all of them emergent.

What about neural systems? A similar analysis provides the conceptual
resources to capture the theoretical shift in neuroscience described and
prescribed by Sporns. Neural synchronization, the critical state, and other
networks can already be explained, at least partially, as witnessed by Sporns's
own research. Furthermore, brain networks and their powers are neither
novel nor "irreducible" to individual interacting parts. What else constitutes
a circuit, in addition to a large aggregation of neurons, their additive and
non-additive interactions, and background conditions? The instructive
lesson imparted by contemporary neuroscience is that a complete description of brain function in terms of individual neurons and their relations is
not required for brain systems to do the explanatory heavy lifting. The reason
is that the core elements and processes that explain the workings of the brain
are difficult to describe at lower levels of organization, and thus require a shift
to a broader perspective. These macro-building blocks are, indeed, constituted by these neurons. But such configurations are stable and autonomous
enough to be identified and represented as wholes, without breaking them
down further to more fundamental units. In sum, the emergence of neural
states resides in the role of complex circuits in producing and maintaining
neuro-psychological states, which does not require a detailed account of how
these processes are physically realized at the unitary level. Thus framed, these
circuits can be effectively black-boxed. It is their status as black boxes that
makes them emergent.

I conclude this array of illustrations by briefly noting that my proposal
captures the essence of many other traditional examples of emergence.
Hempel and Oppenheim were correct, of course, that life and consciousness
are phenomena for which we could not fully account. Unfortunately, despite
scientific strides accomplished during the fifty years or so since the publication of their essay, we still cannot claim to have completely uncovered the
material underpinnings of these properties. At the same time, our incapacity
to explain qualia has not frustrated attempts by psychologists and cognitive
scientists to study conscious mental states, any more than the elusive mystery of life has prevented biologists from studying, systematizing, and categorizing living organisms. Life and consciousness are emergent because

fully explaining these concepts is not a prerequisite for fruitfully appealing to them across the sciences and the humanities. Now, surely, unraveling the nature of life and mind would be a momentous theoretical and empirical achievement. Nevertheless, in the meantime, these concepts can continue to be assumed and employed as black boxes within biological, psychological, philosophical, and other kinds of explanations.

Analogous treatments, I contend, shed light on the alleged emergence of many other classic candidates, like the transparency and liquidity of water, the behavior of markets and other macro-economic phenomena, and various normative properties, such as ethical and aesthetic ones. If these should count as "emergent" at all, we can explain their emergence in terms of black boxes: they are placeholders in causal explanations represented in models.

§8.5. Refinements and Implications

Section 8.4 outlined my emergents *qua* black boxes account and illustrated it with various traditional and more cutting-edge examples. This section refines the view by addressing three implications and a potential objection.

Beginning with the first implication, it follows from our discussion that there are various distinct families of emergent properties. More specifically, emergents can be arranged along two orthogonal dimensions. On the one hand, emergents can be distinguished by the rationale for omitting detail: necessity vs. pragmatic convenience. On the other hand, emergents can be categorized depending on whether they function as explananda or explanantia, that is, whether they play the role of frames or difference-makers relative to an explanatory context. Let me consider both cases, in turn.

First, the micro-structure of an emergent property may be left out because it is currently unavailable or lacking *tout court*. The situation is exemplified by Hempel and Oppenheim's treatment of qualia. Subjective mental phenomena cannot (yet) be accounted for at a physical or biochemical level and, for all we know, a satisfactory account may well be beyond the grasp of science. Here, including the required detail is not an option. At the same time, there are also situations in which, while the microstructure is known, it is irrelevant for the explanation and may thus be omitted for the sake of convenience. This is the case with many aspects of isomers, protein folding, and liquidity, where the amount of included detail is a matter of choice. These two motivations for abstraction and idealization—necessity vs. pragmatic convenience—display substantial differences and thus require independent treatments. Nevertheless, both share a common root: the black-boxing of structure that is irrelevant in the context at hand. This is what makes them instances of the same core feature: emergence.

The second distinction between kinds of emergents mirrors a fundamental difference between black boxes. As we saw in Chapter 4, there are two basic types of placeholders. Frames stand in for patterns of behavior in need of explanation. As such, they are explananda. Difference-makers, in contrast, are explanantia standing in for the mechanisms that produce the patterns of behavior in question. If, as I suggested, emergents are black boxes, the same distinction will apply to these systems and properties. Indeed, examples of both kinds are easy to find. Qualia are often treated as explananda, as they are commonly employed in the philosophy of mind: what underlies my sensation of peacefulness when I gaze at the ocean? Or they can play the role of explanantia, as when I attribute my unwillingness to come to work to a headache. In both cases, the quale is a black box, as witnessed by the lack of structural detail that does not affect the autonomy of the higher-level explanation, when represented in the appropriate model.

In sum, we can classify emergents along two orthogonal dimensions: necessity vs. convenience, and explanantia vs. explananda (Table 8.1). What makes all of them "emergent" is their status as black boxes.

With all of this in mind, let us move on to a second implication of my proposal. Emergence is traditionally presented by contrasting it to reduction (§8.1). Reductionists argue that scientific explanations are invariably deepened by breaking them down to micro-constituents and their relations. Antireductionists retort by endorsing a form of epistemic autonomy, which denies that the reductionist tenet generalizes across the board. Emergence is typically viewed as a way to cash out the antireductionist thesis: higher-level events cannot reduce to micro-levels because they are emergent. This identification of emergence with irreducibility, I have argued, is ineffective. The notion of (anti)reduction is no clearer than the concept that it purports to

Table 8.1 Four Types of Emergents

	Necessity	Convenience
Frame	Type 1	Type 2
Difference-maker	Type 3	Type 4

elucidate, namely, emergence. Thus, explicating the former in terms of the latter is unlikely to yield any groundbreaking insight.[12]

Our discussion provides an alternative. The analysis of emergents as black boxes is consistent with both reductionist and antireductionist outlooks. As such, it should not be contrasted with either. Allow me to elaborate. Modest reductionists acknowledge that higher-level explanations are bona fide explanations. Their contention is that opening the black box and unpacking the details from the abstract and idealized depiction will increase their depth. Sophisticated antireductionists, in contrast, maintain that the "gory details" do not necessarily enhance the explanatory power. I shall postpone a discussion of how the debate can be recast in a more productive fashion until Chapter 10. For the time being, the important point is that both parties, reductionists and antireductionists, can converge in recognizing that emergents are black boxes. The disagreement presumably lies in the fate and role of these constructs. Reductionists will take them as provisional epistemic scaffoldings awaiting to be opened and discarded. Antireductionists will likely favor a more robust interpretation of these boxes, according to which these will prove useful even once they have been opened. But both sides can agree on what makes them emergent.

In sum, the existence of emergent properties and emergent systems is independent of the reductionism vs. antireductionism rhetoric. As we shall see in the chapters to follow, there are some important differences between thinkers belonging to the two traditions. Nevertheless, it would be a mistake to insist that emergence be viewed in opposition to reduction.

A third noteworthy implication of my proposal is that emergence is never an absolute state of a system. Because of the initial step of the construction of a black box, the framing stage, the emergent status of a property or system is always relative to a theory, hypothesis, or explanatory context. Consequently, asking whether or not property P is "emergent" *tout court* is an ill-posed question. This is because it is elliptical. P could be emergent in, say, contexts $C_1 \ldots C_n$ and not emergent in contexts $C^*_1 \ldots C^*_n$.

A few examples should elucidate the main insight. Consider protein folding. The structural details of how and why a protein folds are often irrelevant for the macro-explanation of gene expression, as witnessed by the observation that this information is typically idealized or abstracted away in introductory textbooks in genetics and molecular biology. At the same

---

[12] For a related perspective, see Gillet (2016).

time, if the explanandum is reframed as a more specific inquiry, for instance, if the target is a depiction of the precise mechanisms governing a genetic switch, then many cytological components will become relevant to the explanandum. Protein folding is thus emergent relative to the former frame, but not the latter. For this reason, I contend, it is meaningless to ask if this process is emergent in a context-independent fashion.

As a different illustration, take choice in economics. Is it an emergent property? Once again, an adequate answer requires more specificity. Consider the situation from the perspective of neoclassical economics. All that matters here is what an agent chooses and that the pattern of choices satisfies some consistency constraints. The details underlying the choice in question, why the agents choose the way they do, are completely irrelevant to the explanation provided by expected utility theory. Here utility is a black box and, therefore, it is an emergent feature. Things look very different, however, from a psycho-neuro-economic standpoint. In this case, the psychological mechanisms underlying choice are crucial to the explanation of economic behavior, and cannot simply be omitted, idealized, or explained away. In this context, choice is *not* a black box and thus not an emergent property at all. So, is the notion of choice in economics emergent or not? The answer is: it depends. Relative to the neoclassical framework, it is emergent, because it is treated as a black box. Relative to the psycho-neural framework, in contrast, it is not emergent because the details of the underlying mechanisms cannot be omitted. Analogous explications, I contend, may be given for virtually any other higher-level system and property, from Mendel's laws to isomers, from liquidity to qualia, from life to fitness.

I should make it very clear that the context-dependency of emergents does not entail a rampant relativism according to which the emergent status of an entity varies according to the particular epistemic state of individual speakers, or even the same speaker at different times. The relativity of emergents is coarser: it depends on the specifics of the explanation and thus remains constant across epistemic communities. At the same time, opening a black box does not change its emergent status in all the contexts where such details are irrelevant. It would just specify new explanatory frameworks where the phenomenon in question does not count as emergent.

In sum, black-boxing confers to emergence a pragmatic character that allows one to capture the meaning and significance of this important concept across philosophy and the sciences, mirroring the flexibility of epistemic definitions. Yet, following several metaphysical approaches, it also protects the objectivity and independence of emergence from aleatoric changes in knowledge.

Before moving on, let me address a potential objection. Some readers might be concerned with the breadth of the present proposal, which treats a vast array of properties, systems, and processes as emergent. As long as the phenomenon in question is a placeholder that stands in for behaviors or mechanisms represented in a model—that is to say, as long as it is a black box—the present framework will count the system in question as emergent, even if it is fully explainable at lower, more fundamental levels. To illustrate, in the context of an explanation of why Alex believes it is about to rain, well-known physical facts about the atmosphere can be safely dismissed, effectively rendering lightning emergent, even on the assumption that lightning can be fully explained, and thereby reduced, to physical states. Any analysis willing to countenance fully reducible phenomena such as lightning among emergents, the objection runs, borders on triviality.

This worry, I maintain, is a legacy of the old reductionism vs. antireductionism rhetoric, which treats emergence and reduction as incompatible, mutually exclusive notions. This is an assumption that, I've been arguing, should be rejected. From the perspective of black boxes, reduction and emergence are no longer at odds. Thus, while the framework is indeed applicable to virtually any higher-level entity, the relativization to specific explanatory contexts may help mitigate trivialization worries.[13] Some readers may still cringe at this reconceptualization of a traditional philosophical concept, considering it a mischaracterization. Nevertheless, revisiting some deeply entrenched presuppositions about emergence, I retort, is a relatively small price to pay for a framework that promises to be broad and flexible enough to capture and unify appeals to emergence across several domains.

§8.6. Adequacy Conditions and Advantages

This section wraps up our discussion by, first, presenting some adequacy conditions and, subsequently, some advantages and of my proposal.

Bedau (2012) puts forward three useful criteria to assess any account of emergence. First, he contends, the analysis should be clear and explicit, in

---

[13] The virtual ubiquity of emergence has been noted and defended from trivialization concerns by other authors, such as Humphreys (2016).

particular with regard to the interpretation of dependence and autonomy, which lie at the conceptual core of the concept (§8.1). Second, any characterization of emergence worth its salt must be coherent and consistent with everything we know or commonly accept in both philosophy and science. Third, it should be useful and relevant to empirical work. I defend the promise of my proposal by arguing that it fits the bill on all three scores.

First, it should now be clear in what sense emergents are both dependent upon and autonomous from their basal conditions. To be sure, the identification of emergents with black boxes, in and of itself, poses hardly any metaphysical constraint. Indeed, the black-boxing framework described earlier could also be reconciled with forms of Cartesian dualism, according to which entities belonging to different explanatory levels turn out to be different kinds of substances. Nevertheless, my presentation has assumed—and strongly recommends—a naturalist interpretation according to which, in line with modest physicalist interpretations and current understandings of weak emergence, higher-level systems are fully constituted and exhausted by physical entities. In this uncontroversial sense, emergents ontologically depend on their bases. At the same time, regardless of how this token physicalism is cashed out—supervenience, mereology, classical reduction, or something else—emergents are autonomous from an explanatory perspective. As black boxes, in many contexts, their structural details may be idealized or abstracted away.

Some readers may feel that I am merely sweeping the dust under the rug. Emergence is puzzling precisely because it is hard to see *how* a property or system that is ontologically reducible from its base can also be methodologically autonomous. Simply stating *that* they are leaves the account as mysterious as previous suggestions. I agree. I have signed an "I owe you" note. I still have not shown in what sense emergents are both reducible and autonomous. This chapter simply suggests that emergents can be understood as black boxes, which entails a conditional claim: *if* black boxes are both autonomous and dependent, *then* so are emergents. Turning this conditional into an indicative statement will be the task of Chapter 10.

Bedau's second condition demands that the analysis of emergence under scrutiny must not violate any major philosophical or scientific tenet. Hopefully, the previous chapters have done enough to show that the black-boxing strategy is not only compatible with mainstream scientific practice, but actually sheds light on prominent historical episodes. In a nutshell, readers should be convinced by now that nothing spooky is going on here.

The third and final adequacy condition requires definitions of emergence to be useful for and relevant to empirical work. I trust that the flexibility of the emergents *qua* black boxes framework is evidenced by how smoothly it captures the various examples presented in the previous sections. We have discussed classic paradigms of emergence, such as life and qualia. We have explored other well-known cases: liquidity of water and protein folding. And we have examined less familiar, cutting-edge scientific hypotheses, such as the critical state, neural synchronization, and economic utility.

Alas, it is always hard to derive and support a general thesis from a handful of illustrations. Still, I am optimistic that, at the very least, our survey shows how the proposal at hand holds some promise and deserves to be explored in greater detail. To further sweeten the deal, let me conclude our discussion by presenting three advantages of my reframing of emergence.

First, analyzing emergents as black boxes allows one to retain the intuitive analysis of emergence in terms of causal powers, if one wishes to do so, while avoiding well-known pitfalls. As we saw in section 8.2, a well-known problem with powers is that, without a clear-cut principle for identifying and distinguishing these capacities, it becomes susceptible to trivialization worries. A billion snowflakes have the ability to make my roof collapse, whereas no individual snowflake has that capacity. Is this an aggregative property of the individual snowflakes or a collective property of the snow mass? Is this an emergent capacity at all? Does the snow mass have "novel" causal powers that the collection of individual snowflakes lack? Do we lose or gain abilities when shifting from additive wholes to holistic properties?

The black-boxing framework allows us to introduce talk about the causal power of emergents without getting bogged down into these troubling issues. From our perspective, there is no need to determine whether the powers of the snow mass are novel, a proper subset of the collective capacities of individual snowflakes, or something else. This terminological choice ultimately depends on how the explanandum is framed, which is ultimately a pragmatic matter. As long as the snow makes a difference to the collapsing of the roof, and if this difference-maker can be black-boxed by idealizing and abstracting away the structural details of crystallized water molecules, that is sufficient to dub the property or behavior in question "emergent."

A second advantage of the present analysis is that, despite its pragmatic flavor, it can capture more robust metaphysical intuitions. Consider, for instance, a proposal advanced by Barnes (2012), according to which an entity x is emergent, from an ontological perspective, if and only if x is both *fundamental* and *dependent*. Simply put, emergents are characterized as "fundamental" in virtue of being part of the "real, core stuff" in the world. At the same time, the metaphysical dependence of emergents purports to capture their novelty, that is, their existence over and above their basal conditions. Barnes's list of emergents includes minds, living beings, "gunk," and tropes, as well as examples from physics, such as quantum entanglement.

Prima facie, this kind of robust metaphysical account seems flatly incompatible with the diet mechanistic philosophy presented in Chapter 7, which explicitly rejects ontological claims of this ilk. However, the idea of fundamental dependency has a "diet" counterpart that can be used to preserve the main spirit of the proposal. Fundamentals, after all, are prime candidates for entities that can be assumed and employed in the context of an explanation without specifying their nature and structure—in a word, they are "primitives." In short, the framework is consistent with treating fundamentality or other metaphysical concepts as the conceptual heart of many examples of emergence, if one wishes to do so. Sure, this is an epistemic reading of fundamentality, but it is a type of fundamentality nonetheless.

As a different illustration of how the present account can reformulate metaphysical intuitions, consider a recent debate over the notion of emergence in physics. Following Batterman (2002), Morrison (2012) argues that a higher-level state is emergent when it is not predicted and explained from lower-level micro-processes and is independent of them, in the sense that changes to the micro-physical base would not affect the macro-level emergent process. This proposal seems to capture rather well physical examples of emergence, such as superconductivity. Yet, it is not broad enough to cover applications of the concept in fields such as biology, neuroscience, and psychology, where micro-processes do seem to affect the higher level but can still be omitted from their explanations because of ignorance or pragmatic convenience. One response, of course, would be to dismiss many of the examples discussed earlier as bona fide cases of emergence, reserving the label for processes occurring at the fundamental physical levels.[14] Yet, as I stressed in section 8.4, I would rather opt for a more flexible and ecumenical approach. Ontological physical dependence is a kind of emergence, and an important

---

[14] This form of fundamentalism may well be what Morrison has in mind when she claims that "epistemic independence—the fact that we need not appeal to micro-phenomena to explain macro processes—is not sufficient for emergence since it is also a common feature of physical explanations across many systems and levels" (Morrison 2012, p. 161).

one for sure. But there are several other reasons why structural details might not be included in macro-explanations. All these are different, but equally legitimate, cases of emergence.

A third and final advantage of my framework is that it preserves the centrality of prediction and explanation in emergence, while avoiding Kim's trivialization worries and related concerns about emergence being spooky, mysterious, or otherwise problematic. Emergents play an important predictive and explanatory role across scientific and philosophical domains. Still, contrary to both classic and contemporary accounts, the emergence of a property or process is not inextricably tied to obscurity, lack of explanation, non-derivability *sans* simulation, etc. To be clear, emergence can be associated to any of the preceding, as well as to causal powers, a mix of fundamentality and dependence, or various other metaphysical features. Yet, none of these conditions is necessary or sufficient for emergence. Despite substantial ontological and epistemic differences, the common core feature of all these properties, which makes all of them "emergents," is their role as black boxes in explanation.

§8.7. Concluding Remarks

In conclusion, this chapter proposed a reconceptualization of the long-standing debate on emergence by identifying emergents with black boxes. I suggested moving away from the idea that emergence should be characterized in opposition to reduction, which is a legacy of the old reductionism vs. antireductionism debate. I tried to shift the focus toward the explanatory autonomy of black boxes, understood independently of reduction. Some skeptics may view this as little more than a semantic tweak. However, it turns out to be a surprisingly powerful move. In particular, it allows one to hold a middle ground between a rampant monism, which posits a single, once-size-fits-all definition of emergence, and an uncompromising pluralism, which treats emergence as a hodgepodge of loosely related concepts.

As noted earlier, defining emergents as black boxes is hardly the end of the story. I still need to show how black boxes can be both reducible and autonomous, how they escape the old reductionism vs. antireductionism rhetoric. This will be the objective of the concluding Chapter 10. Before doing so, in the next chapter, I want to turn my attention to another central topic in the philosophy of science: the question of scientific progress. For the time being, I hope to have motivated and inspired an overarching and coherent reframing of emergence that, nonetheless, leaves room for—and, indeed, justifies—various metaphysical and epistemic assumptions, which have a long and hallowed place in the history of philosophy.

9

The Fuel of Scientific Progress

> The central *long run* philosophical problem facing people generally
> is how to maintain a belief in progress without a belief in Utopia.
> And I want to argue that that requires change in our conception of
> progress.
>
> —Hilary Putnam, "A Note on Progress," p. 1

§9.1. Introduction

The previous two chapters explored some philosophical implications of the black-boxing strategy laid out in Chapter 5 and illustrated in Chapter 6. Specifically, Chapter 7 spelled out a "diet" approach to mechanisms that shares a general epistemic outlook with the new wave of mechanistic philosophy while mitigating its ontological presuppositions. Chapter 8 proposed the reframing of an old debate by identifying emergent properties with black boxes. The present chapter continues our philosophical exploration by pointing our discussion toward another hallowed topic: scientific progress. Providing the relevant background requires us, once again, to retrace our steps back to the roots of our discipline: the dawn of contemporary philosophy of science.

As we saw in Chapter 1, logical positivism built upon a popular, intuitive, and prima facie compelling account of the scientific enterprise. From this familiar perspective, the main goal of science is to provide a reasonably complete and generally accurate description of the universe. For the most part, positivists maintained, scientists have been successful at attaining this noble endeavor. Successive generations of researchers have filled in more and more details on this grand canvas depicting the universe. Of course, everyone was perfectly aware that this grandiose wall or mosaic-like construction was not nearly complete. Obviously, there is much that science does not know. And parts of the story, as currently narrated, are bound to be mistaken. Still,

Black Boxes. Marco J. Nathan, Oxford University Press. © Oxford University Press 2021.
DOI: 10.1093/oso/9780190095482.003.0009

positivism perceived an overall upward trend toward the attainment of truth or, at the very least, increasingly precise approximations thereof.

Optimistically naïve as it may appear to a contemporary audience, this view of science provided a simple but effective answer to long-standing philosophical quandaries. In particular, it offered a relatively straightforward account of scientific progress, according to which science advances through a slow, constant, and painstaking accumulation of truth. It also ascribed a precise role to philosophy, namely, to make explicit the methodological standards, the tacit canon of rules that underlie this quest for knowledge.

Few readers will be surprised to hear that the good old days of positivism are gone. Once-popular attempts to uncover the structure of theories, the logic of confirmation, and the form of explanation are now scornfully dismissed as ahistorical, shallow oversimplifications. In the wake of Kuhn's groundbreaking work, the positivist ideal was replaced by a historically informed and realistic depiction of scientific theory and practice. Importantly, Kuhn and his allies did not try to replace the formalized accounts of "good" scientific reasoning advanced over the previous decades by authors such as Popper, Carnap, Hempel, Reichenbach, and Nagel with more apt characterizations.[1] The "purely observational" vocabulary on which positivism rested the foundation and ultimate epistemic warrant of the scientific edifice could not be amended and restored via a process of enrichment. The Kuhnian blade cut deeper. It dismissed as misguided the entire project of spelling out an objective, logical, mathematized account of scientific reasoning. The question then became: what should we put in its place?

Over half a century has passed since the first publication of *Structure*, in 1962. Decades of discussion, clarification, and refinement did not yield a fully developed, viable replacement for the model of science presupposed by positivism. Some of Kuhn's most radical colleagues—most notably Paul Feyerabend, Norbert Russell Hanson, and, later, Bruno Latour—were willing to bite the bullet and throw in the towel. They interpreted Kuhn's message as flatly rejecting the existence of a single set of epistemic standards that could objectively capture the rationality of all substantial decisions undertaken

---

[1] As Hacking (1983) aptly notes, Kuhn did not single-handedly engineer this transformation in the history and philosophy of science. When *Structure* was first published in 1962, similar themes were being expressed by a number of voices and the history of science was forming itself as a self-standing discipline. The fundamental transformation in philosophical perspective was that science was finally becoming a *historical* phenomenon.

across and within fields. In Feyerabend's (1975) provocative slogan, when it comes to scientific method and rationality, "anything goes."

Kuhn himself, however, was not quite willing to take it that far. In subsequent publications, he firmly resisted the complete relinquishment of systematic discussions of rationality, in favor of sociological explanations of scientific behavior. In particular, adamantly rejecting charges of "mob psychology," Kuhn insisted on the existence of bona fide epistemic methodological standards which demarcate "good" from "bad" instances of scientific reasoning. And yet, contrary to a core tenet of positivist orthodoxy, these rules are neither algorithmic nor perfectly objective. Consequently, they cannot be articulated generally or made fully explicit. Nevertheless, Kuhn stressed, science is a rational endeavor and must be portrayed as such.

Over the years, there have been remarkable attempts to embed Kuhn's insights into a rigorous framework accounting for scientific rationality and success.[2] However, around the dawn of the new millennium, mainstream philosophy of science underwent a paradigm shift of its own. The field eventually abandoned the project of developing a grand overarching account of the scientific enterprise. New approaches traded in the search for generality for a focus on particular sub-disciplines: physics, biology, neuropsychology, and the like. As a result, Kuhn's expensive tab is still largely unsettled.

The systematic development of a post-Kuhnian replacement for the outmoded positivist image of science remains a prominent item on philosophy's unfinished agenda. This is a substantial endeavor that transcends my present concerns. Still, my account of black boxes has something to offer to the overall project. My aim, in the pages to follow, is to show that the three-step recipe for the construction of black boxes, developed in the first half of the book, sheds light on an issue that lies at the core of any general account of science, namely, progress.

Here is an outline of the current chapter. To kick things off, section 9.2 reconstructs the genealogy of the concept that, more than any other, has sparked controversy by undermining traditional accounts of the advancement of science: meaning incommensurability. Section 9.3 surveys the strategy adopted by many contemporary philosophers to deal with the puzzle, namely, proposing and refining referential models of meaning. The remaining sections bring black-boxing into the equation. Readers well-versed in these philosophical debates should consider beginning here. Specifically, section 9.4 shows how

---

[2] See, for instance, Lakatos (1970); Shapere (1974); Laudan (1977); and Kitcher (1993).

black boxes can be used to enrich the conception of progress underlying referential models, while mitigating the perilous consequences of meaning holism. Section 9.5 borrows some insights from referential accounts of theory-change to sharpen and develop the three-part recipe presented in Chapter 5. Finally, section 9.6 wraps up our discussion of scientific progress with a few concluding remarks.

Before getting down to business, allow me to preempt a potential objection.[3] Some readers may take my call to revisit traditional questions of progress, incommensurability, sense, and reference as a step in the wrong direction. These are topics reminiscent of 1970s philosophy of science which, like most of philosophy back then, took its cue from the philosophy of language. Many have come to view the divorce from the issue of language, and its replacement with an increased attention to empirical practice, as one of the great achievements of contemporary philosophy of science. This being so, should we really resurrect these tired old issues of meaning? Even philosophers of language have finally set them aside, and for good reason.

To be clear, I have no intention of dragging philosophy of science back to its positivistic roots. Nor do I want to disavow the current naturalistic stance, as my attention to various case studies will hopefully testify. Nevertheless, the question of progress is a central constituent of any account of science worth its salt. And to the best of my knowledge, it has not been solved. It has been quietly swept under the rug, where it has lain dormant ever since. My goal is hardly to rehash the good old linguistic turn. It is to bring important philosophical issues back onto the main stage. And if an honest, "naturalized" discussion of progress requires reviving long lost concepts such as meaning, reference, and incommensurability—is there a better alternative currently on the table? —then so be it. Let's not throw out the baby with the bath water. If you just can't stomach it, flip to the following chapter. No hurt feelings!

§9.2. The Roots of Incommensurability

Let me begin by setting things straight. Section 9.1 casually referred to "the" logical positivist account of science. Strictly speaking, this is an oversimplification. There admittedly is no uniform, monolithic depiction of science collectively adopted by all positivists. Schlick, Carnap, Hempel, Neurath,

---

[3] I am grateful to John Bickle for lucidly bringing this reaction to my attention.

and Reichenbach—just to mention a few key figures—had subtly different conceptions of the scientific enterprise. Furthermore, Kuhn's proposal was equally at odds with the views of Karl Popper, who would have vehemently rejected any association with positivism.[4] Nevertheless, setting aside substantial disagreement among the aforementioned authors, the main protagonists of philosophy of science during the first half of the twentieth century did share a general outlook. As Hacking (1983) observes, this common model presupposes the following assumptions. First, a sharp distinction is drawn between observation and theory and between the context of discovery and the content of justification. Second, science is largely a cumulative endeavor. Third, there is a general methodology that unifies all of science. Fourth and finally, this underlying methodology can be characterized precisely in a formal and essentially ahistorical fashion. For lack of a better term, I call this background conception the "positivist" image of science.

This cumulative approach to scientific knowledge is epitomized in Nagel's *The Structure of Science* which, as discussed in Chapter 2, has become a classic of late positivist philosophy. Nagel tells an optimistic tale of stable structure and continuity across the epochs. Still, he also realized that, from time to time, theories are in dire need of a makeover or even a replacement.

This commonplace observation raises a crucial question. When is it rational to switch from an older theory T_1 to a newer theory T_2? Nagel, who absorbed the positivist lesson well, had a simple and effective answer. For such a switch to be warranted, the new theory T_2 must be able to explain all the phenomena already accounted for by T_1 and, similarly, preserve all the successful predictions of its predecessor. In addition, T_2 should exclude parts of T_1 which turn out to be faulty, erroneous, or otherwise misguided. Alternatively, the new theory must cover a wider range of explanations and predictions compared to the old one. Ideally, the new theory accomplishes both goals, in which case, Nagel claims, we can say that T_2 "subsumes" T_1.

This rough sketch corresponds to the broad picture rejected by Kuhn and his allies. Specifically, Kuhn famously replaced this standard monolithic model with a two-factor approach, which distinguishes two stages of science: *normal science* and *revolutionary science*. In a nutshell, normal science is a puzzle-solving activity largely engaged in relatively minor tinkering

---

[4] This is chiefly because Popper rejected "bottom-up" theories of confirmation, a hallmark of positivism, which purport to build knowledge from individual observation statements. Popper espoused a "top-down" approach, which eschews all forms of induction and confirmation in favor of a hypothetico-deductive falsificationist scientific methodology.

with established theory. Against the assumptions of positivists and their traditional adversaries, confirmation, verification, and falsification play a relatively marginal role in everyday "normal" scientific practice. For instance, contrary to Popper's dictum, mismatches between theory and observation should not be considered falsifications. Rather, they are routinely treated as "anomalies," that is, nagging counterexamples to be captured and explained away by the reigning paradigm. Nevertheless, despite these substantial differences, Kuhn's normal science can be easily reconciled with Nagel's model. In particular, normal science may be viewed as a cumulative collection of statements and concepts within a specific domain. Progress is tantamount to any contribution to this growing body of knowledge. Yet, this is only one half of Kuhn's story and not the most exciting one.

Kuhn observed that sometimes anomalies stubbornly resist resolution. Rather than washing away, they pile up, and a few may come to be viewed as especially pressing. As researchers try to fix the problem, counterexamples accumulate. Consequently, the field enters a state of *crisis*. The typical way out of this quagmire is the development of a fresh start, equipped with a host of novel tools and concepts. As the new framework rapidly establishes itself and makes progress, superseded questions and ideas are set aside and, eventually, they are forgotten. When this happens, when an older paradigm is replaced by a newer one, we have what Kuhn dubs a *scientific revolution*.[5] The new paradigm eventually crystallizes into normal science, adopting a consensus-forging role, producing fresh anomalies, which eventually trigger a crisis, followed by another revolution, and so on, in a continuous cycle.

The outcome of the revolution—the new paradigm—typically exhibits different goals and interests compared to its predecessor. The revamped normal science may ask innovative questions, postulate novel concepts, posit distinctive laws, and so forth. Still, the occurrence of a revolution, by itself, does not jeopardize the overall rational or progressive trajectory of science. The threat to progress and rationality stems from the very nature of paradigm shifts, which Kuhn famously compares to religious conversions and gestalt switches in psychology. Members of the new paradigm, he claims, "live in a different world" from their predecessors. They speak different languages

---

[5] Providing a precise, univocal characterization of "scientific paradigms" is a notoriously thorny endeavor. Following Hacking (1983), I focus on two prominent meanings. First, paradigms can be considered achievements, in the sense of being the ultimate goal of normal science. Second, paradigms are sets of shared values: collections of common methods, standards, and basic assumptions that jointly underlie specific lines of inquiry.

that cannot strictly be compared or translated into each other. The only way to transition from one to the other is via a *eureka*-style intuition, as opposed to deliberative reasoning. This is where trouble begins. "Living in a different world" has substantial implications for progress.

On Kuhn's picture, *sensu stricto*, where novel paradigms carry new languages with them, one might not even be able to convey the ideas of the replaced theory in the language of the replacing one. Indeed, Kuhn initially suggested that there is literally no way of specifying a theory-neutral language in which to express and compare the two frameworks. This is a striking departure from the old cumulative model of scientific knowledge. Recall that Nagel did recognize the need for theory change. Still, despite the occasional setback or difference in focus or perspective, the new theory always takes the success of its predecessor under its wings, while eschewing some problems, failures, and misconceptions. The basic idea is that the two theories can be rationally compared, and the "better" one is selected. After all, substantial change in theory is warranted if and only if the new paradigm explains the known data and predicts new observations more accurately than its predecessor. This, Kuhn claims, is what happens in the context of normal science. Yet, he continues, when we zoom out and focus on larger time scales, science does not work that way. After a revolution has swept in, a substantial component of the old paradigm is dismissed and forgotten. It eventually becomes accessible only to historians who, through slow and painstaking work, are able to reconstruct the discarded *Weltanschauung*.

This, simply put, is Kuhn's notorious concept of *incommensurability*, which leads to conceptual relativism, the controversial doctrine that the language used in a field of science changes so radically during a revolution that the old and the new theories are not mutually inter-translatable.[6] Etymologically, "incommensurability" derives from ancient Greek mathematics, where the term had a precise meaning, denoting two lengths which have no common measure.[7] Kuhn borrows this idea and applies it, figuratively, to the comparison of scientific paradigms. The notion of incommensurability may be used to describe different phenomena, with various controversial implications.

---

[6] As Kitcher (1978) remarks, it is interesting that Kuhn and Feyerabend, vigorous contenders of the relevance of history for the philosophy of science, have also advanced theses which imply that the task of the historian of science cannot be successfully completed.

[7] Two lengths p and q have a common measure if it is possible to lay x of one against exactly y of the other, thereby measuring p in terms of q, or vice versa. Not all lengths are commensurable, as shockingly discovered by the Pythagoreans when they realized that the diagonal of a square is "incommensurable" to the length of the sides.

Following Hacking, I distinguish three distinct concepts of incommensurability, as they pertain to the history and philosophy of science.

First, there is what can be called "topic incommensurability." As noted, Nagel followed his fellow positivists in claiming that when a theory is replaced by a successor, the new theory always does the same job "better" than the old one. Kuhn and Feyerabend saw that this view is unduly narrow. The paradigm that is established in the wake of a revolution is frequently quite different from its predecessor. For instance, it may ignore a lot of past successes and it might re-describe phenomena in different ways. When this happens, later generations may find previous theories unintelligible, until they play the role of historians, effectively reinterpreting the old theory. Revolutionary as it was in the 1960s, this topic-incommensurability has now become widely accepted. Few contemporary scholars would deny that, while subsumption captures many prominent episodes in the history of science, it does not cover all cases.

A second, stronger kind of incommensurability, which is central to much of Feyerabend's provocative discussion in *Against Method*, is what Hacking dubs "dissociation." This occurs when the gap between a superseded theory T_1 and its successor T_2 is so significant that, from the perspective of T_2, one cannot even attach a truth value to many statements of T_1. Hacking illustrates this phenomenon with the example of Paracelsus, a sixteenth-century alchemist whose theories of matter were so farfetched from the perspective of contemporary chemistry that they cannot be matched with anything we want to state today. Sure, one may express Paracelsus's claims in modern English. Yet, we often cannot meaningfully assert or deny what is being said.

Hacking's third kind of incommensurability stems from a puzzle concerning the meaning of theoretical terms, that is, terms that refer to unobservable states (cf. §6.4, Chapter 6). The problem is simple but deep. Children and adults alike often learn the meaning of mundane concepts by *ostension*. To illustrate, one can convey the meaning of "apple" by displaying a paradigmatic apple. Similarly, one can acquire the word "bitter" by tasting a substance that falls under said category. But how do we assign meaning to concepts referring to entities that cannot be observed? How do we learn about electrons, black holes, molecular chaperones, protons, and so forth? One obvious suggestion is that we acquire these concepts by memorizing definitions. Upon further scrutiny, however, this strategy turns out to be a non-starter. For one, precise definitions are hard to come by. Furthermore, and even more problematic, the definitions of many theoretical terms appeal to other, even more technical notions. For instance, learning the standard definition of "electron" as a negatively charged subatomic particle presupposes that the subject already grasps the concept of charge, which is arguably more complex and theory-laden than "electron" itself. These and similar difficulties suggest a different, more plausible alternative. We explain theoretical terms by specifying a theory. The meaning of words like "electron" and "black hole" is provided by their position within the structure of the entire corresponding theory. From this holistic standpoint—pioneered by Duhem, endorsed by many logical positivists, and developed to its fullest extent by Quine[8]—it follows that, say, "mass" does not mean the same thing in classical and relativistic physics, and "planet" is a different concept for Ptolemy and Copernicus. The reason should now be evident. If the meaning of a theoretical term is determined by an entire corresponding theory, changing the theory thereby changes the meaning of all its relevant concepts.

At first blush, these conclusions will strike some readers as inconsequential and, perhaps, quite plausible. Did Einstein's groundbreaking intuitions not change the very meaning of "mass" by relativizing it to a frame of reference? Meanings are constantly in flux. All this is fine. Yet, problems arise as soon as we try to compare theories. If terms like "electron" and "mass" get their meaning from their place within a network of statements and laws, then, as noted, when the theory is modified, the meaning of these terms changes as well. But if the meaning of core concepts varies from theory to theory, how can we compare the theories themselves? If, in principle, theories never talk about the same thing, then we have no common measure to assess them. Hence arise the notorious problems of incommensurability and theory change, which turn Nagel's plausible doctrine of subsumption, as well as the very possibility of "crucial" tie-breaking scientific experiments, into a logical impossibility. In what follows, it is this third and strongest *meaning-incommensurability* that will be the focus of our attention.

The thesis of meaning-incommensurability was met with cries of outrage. Some dubbed it untenable because its rests on the fundamentally incoherent idea of incompatible conceptual schemes. Others rejected it on the grounds that there is clearly enough similarity of concepts across

---

[8] See Duhem (1954); Carnap (1956a); Hempel (1965, 1966); and Quine (1953).

successive theories to allow for meaningful comparisons.[9] Kuhn himself was taken aback by the way his work, and the contributions of his colleagues, produced a crisis of the concept of rationality in the philosophy of science.[10]

In the wake of the heated debates ensuing the publication of *Structure*, Kuhn (1977) clarified his position by stating that he never intended to deny the customary "virtues" of scientific theories. Theories should be accurate, logically consistent (both "internally," in their own structure, and "externally," when conjoined with other theories), broad in scope and rich in consequences, simple in structure, and fruitful. Following these observations, Kuhn's own "revolution" in the philosophy of science may thus be conceived along two orthogonal axes. First, these and similar values are never sufficient to make a decisive choice among competing theories. Other considerations come into play, factors for which there could, in principle, be no formal decision-algorithm. Second, Kuhn maintains that proponents of rival theories are like native speakers of different languages. There are substantial limits to what can be effectively conveyed. Still, despite these communication breakdowns, scientists immersed in different theories can, albeit with difficulty, present to each other concrete technical results, available to those who practice within each theory, and their implications.

In conclusion, *pace* authors like Feyerabend and Shapere, who would like to expunge semantic issues from the philosophy of science, the notion of incommensurability raises, at the most basic level, a question concerning how theoretical concepts acquire their significance. And addressing this quandary, unfortunately, presupposes at least a rough conception of meaning—a notoriously thorny endeavor. There is, however, some good news. The past few decades have witnessed the development of *referential models of meaning*, which promise to resolve the problem of incommensurability by allowing radically different theories to talk about the same entities. Section 9.3 turns to some popular variants of this influential approach.

---

[9] The first critique was famously explored by Davidson (1974). The second route was developed primarily by Shapere (1966); Achinstein (1968); and Kordig (1971).

[10] Hacking (1983) suggests that Kuhn did not originally intend to address the issue of rationality at all. Things are different, however, in the case of Feyerabend, a philosopher whose radical ideas often overlap substantially with Kuhn. Feyerabend is a longtime foe of dogmatic rationality. For him, there is no canon of rationality, no privileged class of "good reasons," and no absolute paradigm of science. There are many rationalities, as opposed to a single one, and the choice among them can never be fully objective.

§9.3. Referential Models and Incommensurability

Traditional philosophical discussions of language are heavily influenced by the seminal work of Frege, who distinguished between two components of meaning: the *sense* of an expression, which must be grasped by all communicators, and the *reference*, the entity denoted by the expression in question. Now, supplement Frege's insight with the holistic presupposition—which came to be accepted by logical positivists but, to be clear, Frege himself would have rejected—that the sense of a term is determined by its place within a network of theoretical statements. It follows that the meaning of the term always shifts whenever the underlying theory undergoes change. *Ecce* meaning incommensurability, as presented and discussed in section 9.2.

Putnam pioneered a response to the problem by enriching Frege's picture.[11] Meaning, he claims, has not two, but four components. First, each expression has *syntactic markers*, which characterize its syntactic structure (e.g., noun vs. adjective vs. verb). Second, words have *semantic markers* that assign the term in question to the appropriate category. Thus "rabbit" denotes organisms and "iron" picks out a metal. Putnam's most original contribution is the third constituent of meaning, the *stereotype*, which can be characterized roughly as the conventional image or description associated with a word. For instance, rabbits are typically depicted as furry, timid, fuzzy mammals. Similarly, water is associated with an image of a tasteless, transparent, thirst-quenching liquid. Stereotypes can be inadequate. Some, even all, elements of the description may be mistaken, as revealed by the observation that not all rabbits have long ears and, alas, not all water comes in clear, tasteless form. Furthermore, conventional depictions are subject to change. The once-prominent stereotype of wolves *qua* vicious, blood-thirsty beasts is fortunately turning into a thing of the past.

If syntax and semantic markers are too general, and stereotypes are subject to change, is there anything specifically tied to an expression that remains constant in meaning across contexts? The answer lies in the fourth component of meaning: *reference*. While stereotypes may evolve as we find out more about a certain thing or kind, the denoted entity—or the extension, the set of entities falling under the predicate, in the case of properties and natural

---

[11] Frege's classic distinction between *Sinn* and *Bedeutung* comes from his 1892 article. Putnam's well-known proposal is detailed in "The Meaning of Meaning" (1975b).

kinds—does not vary. Thus, the fundamental identity for expressions is determined neither by senses nor by stereotypes. It comes from reference.

With this in mind, it becomes clear how it is possible for concepts embedded in radically different theories to nevertheless talk about the same things. This happens when a term maintains the same stable reference across Kuhnian paradigm shifts. Thus, Democritus, Newton, Laplace, Thomson, Lorentz, Bohr, and Millikan, all of whom have very different theoretical presuppositions concerning physical particles, can be talking about the same entities when they use the term "electron." In short, this referential approach provides the resources to address the meaning of theoretical terms without being lured into problems—or pseudo-problems—of incommensurability and relativism.

Things, unfortunately, are not that simple. Putnam's sketch works reasonably well for "success stories," where authors with very different presuppositions may nonetheless use a term to refer to the same natural kind: electrons, viruses, gravity, and the like. But how do we apply this model to concepts, such as *acid*, which have competing definitions and are very likely not to be homogeneous natural kinds? Even more problematically, proponents of theories positing nonexistent entities, such as *aether* and *caloric*, seemingly communicate their ideas just as well as researchers with different views about "real" theoretical entities, like viruses and electrons. How do we account for this observation? How do we explain the substantial contribution of these theories to contemporary science, despite the lack of stable referential relations across the board? Some dismiss these examples on the grounds that "the notion of meaning is ill-adapted to philosophy of science. We should worry about kinds of acids, not kinds of meaning" (Hacking 1983, p. 85). Others have picked up the tab by revising Putnam's insights.

An influential refinement has been offered by Kitcher.[12] Situating himself within a philosophical tradition which feels uncomfortable with intensional entities like Fregean senses, Kitcher urges the benefits of a "referential" approach to the semantics of scientific terms. As noted, the development of an extensionalist (referential) account of conceptual change was already well on its way, at the time Kitcher was writing.[13] However, Kitcher argues, referential change, in and of itself, is neither necessary nor sufficient for conceptual

---

[12] Kitcher develops his views over a series of publications. The following reconstruction brings together aspects from Kitcher (1978, 1982, and 1993).

[13] The idea of cashing out incommensurability in terms of reference was an old adage, proposed over a decade earlier by Scheffler (1967), and refined by Putnam, among others.

relativism.[14] What would truly inhibit inter-paradigm understanding is a particular kind of referential change, culminating in a mutual inability to specify the referents of terms used in elucidating the rival position. Thus, at the heart of relativism is the thesis that scientists working in the same fields across revolutions are unable to compare their theories. This could happen when, for some expression-types, we are unable to specify a set of entities whose referent belongs to that set. In his own words, "the idea of the incomparability ('incommensurability') of two scientific theories presupposes that there is *no* adequate translation of the language used in presenting one of those theories in the language of the other" (Kitcher 1978, p. 529). What exactly is going on here?

In the years following the publication of *Structure*, philosophers of science typically responded to the challenge advanced by Kuhn and Feyerabend— the issue of incommensurability presented in section 9.2—by presupposing, more or less explicitly, that the only alternative to conceptual relativism is the claim that all scientific language can be translated via a context-insensitive theory of reference. This, Kitcher argues, is a false dichotomy. Neither relativism nor traditional invariant approaches offer an adequate account of how scientific terms evolve across paradigms.

Kitcher unfolds his argument by focusing on a detailed case study: the language developed by Joseph Priestley, Henry Cavendish, and other advocates of phlogiston theory. Such theory, first introduced in section 4.4 of Chapter 4, purported to provide an explanation of various chemical reactions and, in particular, the process of combustion. The problem, simply stated, is that the concept of phlogiston was intended to pick out a substance—a "principle"— which is emitted in all instances of combustion. Yet, since there is no substance which is emitted in all cases of combustion, the term "phlogiston" fails to refer. This being the case, given that "phlogiston" is a non-referential concept, then how could Priestley and his colleagues have made the significant discoveries which legitimize their rightful spot in the Pantheon of scientists?

Let me clarify the issue by restating it in the form of a dilemma. On the one hand, in rejecting the existence of phlogiston, we are forced to concede that not only "phlogiston," but also all other derivative notions that contain "phlogiston" as a constituent—such as "phlogisticated air," and "dephlogisticated

---

[14] A trivial kind of conceptual relativism *sans* reference change occurs when languages contain non-overlapping expressions. More interestingly, reference-change may occur without incommensurability, if shifts in reference can be specified in the new language.

air"—are non-referential. This entails that statements involving these entities are false, as the required truth-makers do not exist. On the other hand, it is hard to deny that when Priestley talks about, say, the positive sensation and beneficial effects of breathing dephlogisticated air, he is not speaking falsely at all. Indeed, he seems to be making true statements about a substance still unknown to him, namely, oxygen. But how is this possible? How can a statement be true when one of its core constituents fails to refer? How can terms like "phlogiston" and derivative notions be both non-referential and explicative?

Note that we cannot resolve the conundrum by appealing directly to Putnam's insights. This is because reference here clearly does *not* remain constant across theories. As far as we know, "oxygen" picks out a natural kind, a chemical element with atomic number 8. "Phlogiston," in contrast, has no real-world counterpart. The solution, Kitcher suggests, is not to settle for an inferior translation that renders many of Priestley's statements inexplicable from our contemporary perspective. Rather, he claims, we should abandon the search for an overly demanding context-independent theory of reference in favor of a context-sensitive one. From this perspective, different tokens of, say, "dephlogisticated air" pick out different entities. Some occurrences refer to oxygen, some to hydrogen, some fail to refer, etc.

How can tokens of the same expression-type denote different things? The key to answering this question comes from the philosophy of language. In the 1970s, it became clear that, contrary to the then-received view, it is possible to use a term referentially without being able to offer any informative description of the term itself.[15] The reason is that usage of words is parasitic on the presence of experts within a linguistic community. These are specialists who are able to make fine-grained distinctions, real or apparent, between the extensions of similar concepts, without necessarily having precise definitions in mind. This notion of direct reference is fixed via a causal story beginning with some primal act of baptism.[16] Important as it is, this social dimension of language, Kitcher maintains, should not lead us to disregard the crucial role of speakers' *intentions* in determining the reference of expressions.

---

[15] Major contributions are Kripke (1972); Donnellan (1970, 1974); and Putnam (1973).

[16] More precisely, following Kripke and Donnellan, Kitcher assumes that this process of fixing reference takes the form of a "historical explanation." The referent of a token is that entity which figures in the appropriate way in a correct explanation of the production of the token in question. Any such explanation will consist in a direct or indirect description of a sequence of events that begins with a primal act of baptism and whose terminal member is the production of the token.

Within scientific communities, Kitcher views linguistic practice as governed by three general principles. First, a maxim of *conformity*: refer to that to which others refer. Second, a maxim of *naturalism*: refer to natural kinds. Third, a maxim of *clarity*: refer to that which you can specify. Under many circumstances, these principles may come into conflict. To illustrate, consider a physicist who has discovered what appears to be a new kind of particle. Should this researcher try to refer to the same entities that have been theorized by others before? Or should they posit the existence of an altogether new natural kind? What if they are unable to specify the nature of the newly identified entity with clarity? As a result, scientists have to choose, more or less consciously or deliberately, among them. And since on different occasions even the same author may opt for different choices, different tokens of the same expression-type may refer differently.

How do we determine the reference of each token of an expression, given this context-dependency? The correct hypothesis about the reference of a speaker's token, Kitcher suggests, is the hypothesis which best explains why the speaker said what they did. Positing a *principle of humanity*, which enjoins us to impute to the author a pattern of relations among psychological states and the world that is as similar to ours as possible, the best explanation will tell the correct story about the speaker's intentions in uttering a term.[17] This involves relating those intentions to the external circumstances of the speech act and the speaker's verbal behavior, with the understanding that, on different occasions, even the same author can be guided by different maxims. Good. But how does all of this work in practice?

Scientific expressions are associated with a complex conceptual apparatus that Kitcher calls their *reference potential*. The reference potential of an expression-type is a compendium of the ways in which the denotation of term-tokens is fixed for members of a community. More precisely, the reference potential of a term, relative to a speaker, is constituted by the set of events which the speaker is disposed to admit as initiating events of tokens of that term. In turn, community members are defined as individuals who are disposed to accept the same initiating events for tokens of a term in virtue of sharing a set of dispositions to refer in various ways.

Why do scientific terms frequently have heterogeneous reference potentials, that is, reference potentials which include a broad and diverse

---

[17] This "principle of humanity," which Kitcher borrows from Richard Grandy (1973), is also known in philosophy as the "principle of charity."

set of initiating events? The reason is that scientists who engage in different projects frequently find it useful to initiate their tokens by different events. For instance, these tokens may correspond to the different ways in which a chemical substance or reaction can be produced. When this occurs, as it frequently does, we may reasonably conclude that the term in question is "theory laden."

What exactly is Kitcher alluding to? Setting technicalities aside, a simple example should help drive the point home. Consider the concept of gene, in the context of Mendelian genetics. When classical geneticists talked about "genes," what exactly were they referring to? The obvious answer is that genes corresponded to *units of function*. The effects of the genotype of an organism on their phenotype, from this standpoint, could be traced back to the effect of a gene or set of genes. This is true. However, it can hardly be the entire story. Within the theory of classical Mendelian genetics, genes also played at least two other theoretical roles.[18] Genes were also taken to be *units of recombination*, in the sense that changes in linkage relationships either separated genes that were previously linked or linked genes which previously segregated independently. Finally, genes were characterized as the *units of mutation*: changes in genes give rise to new alleles, variants of that same gene. In short, as we shall discuss in greater detail in the following section, scientists within and across eras and fields can use the same term, "gene," to pick out very different entities. This can be captured effectively by claiming that "gene" is a theory-laden term with a multifaceted reference potential. What makes classical geneticists a "linguistic community" is their shared disposition to use the term "gene" to refer to the same entities or events.

The thesis that theory-laden scientific expressions have a heterogeneous reference potential—that is, these terms may refer to different entities in different contexts—is the conceptual core of Kitcher's extensionalist solution to both incommensurability and scientific progress. I take up both issues, in turn.

Begin by focusing on the former issue: incommensurability. The heterogeneous reference potential of an expression depends, at least in part, on the theoretical presuppositions which pertain to the particular paradigm to which the expression belongs. Hence, it will often occur that the reference potential of a term belonging to the language of a theory cannot be matched by any expression in its post-revolutionary successors, when these

---

[18] For an insightful discussion of these theoretical roles, see Griffiths and Stotz (2013).

subsequent theories do not share these same presuppositions. If by "incommensurability" Kuhn and Feyerabend simply mean a mismatch of reference potentials across theories, then there are indeed cases of conceptual relativism in the history of science. Priestley's phlogiston is a prime candidate for an expression whose reference potential lacks an analog in Lavoisier's chemistry. However, this type of relativism does not involve any inability to specify the referents of tokens across paradigm shifts. In short, reference potential allows for radical conceptual revision without conceptual discontinuity. Theory-ladenness does not thwart the possibility of successful communication. In fact, it is an integral component of it.

Moving on to the second point—progress—reference potential also lies at the heart of Kitcher's account of the advancement of science, which revolves around two types of cognitive progress: *conceptual* progress and *explanatory* progress.[19] In conceiving of science as "progressive," we envision it as a sequence of consensus practices that get better and better with time. Improvement need not be constant, but there should be a general upward, positive trend. The challenge, of course, is to clarify what "better and better" and "upward" mean in this context, given that the mere accumulation of facts is off the table. Conceptual progress can now be understood as progressively adjusting the boundaries of our categories and providing more adequate specifications of referents, in a way that makes them conform to natural kinds. These kinds of shifts can be captured as changes in the reference potential of key theoretical terms. Explanatory progress, in contrast, consists in improving our understanding of the structure of nature, an account embodied in the schemata of our practices. Improvement involves either matching our schemata to the mind-independent ordering of phenomena or producing schemata that are better able to meet some criterion of organization, such as greater unification. Progress, in short, in not merely an increment in knowledge. It also encompasses better, more accurate, more perspicuous concepts, better equipped to describe and explain nature.

In sum, Kitcher's discussion of reference potential provides a powerful tool to analyze, address, and explain away meaning incommensurability, the core of Kuhn's and Feyerabend's critique of traditional philosophy of science,

---

[19] "Cognitive" progress is understood in contrast to "practical" progress, which involves technological advancements and other means of manipulating nature. While cognitive progress may initially appear to be more ethereal and elusive, Kitcher (1993) argues that it can be approached far more easily because of the relative narrowness of the set of human impersonal epistemic goals compared to the constantly in-flux practical goals.

and to capture the notion of progress. Incommensurability boils down to mismatch in reference potential. This form of conceptual relativism is widespread in the history of science. But it does not undermine the possibility of viewing the relation between theories as "progressive."

With this in mind, it is time to bring black boxes back on the main stage. In the remainder of the chapter, I articulate and defend two related claims. First, the black-boxing strategy, detailed in Chapter 5, brings to the table a significant array of resources that can be used to supplement and refine referential accounts of conceptual change. In particular, it suggests a richer and more plausible model of progress—a long-standing and significant philosophical issue that has been unduly neglected over the last quarter of a century. This thesis will developed in section 9.4. Second, reference potential, explanatory extension, and other notions discussed in the first part of this chapter can be used to further clarify the nature and structure of black boxes. This latter tenet will be the subject of section 9.5.

§9.4. Black Boxes and Reference Potential, Part I

The previous section briefly covered some influential approaches to the time-worn issue of scientific progress. Referential models have two substantial strengths. First, they allow talk about theoretical change without positing senses or other intensional entities that many authors find ontologically suspicious. Second, they pave a way for heeding Kuhn's and Feyerabend's call to take seriously the history of science, while eschewing the more radical and problematic implication of their view: meaning incommensurability.

The strategy outlined in section 9.3 is not devoid of controversial implications. Kuhn himself was sympathetic to Kitcher's proposal that the language of modern chemistry can successfully identify the referents of the key expressions of phlogiston theory. Yet, Kuhn did not fully accept Kitcher's characterization of reference-determination as a bona fide "translation" and the related suggestion to bring talk about incommensurability to a close.

We may paraphrase Kuhn's qualms along the following lines. Referential agreement may well be a necessary condition for comparing theories and ideas across paradigms. But is it sufficient? Kuhn answers in the negative. Communication requires more than a shared interpretation based on extensional semantics. It presupposes a true "translation," that is, agreement on what is said about the referents. Thus, formulations and resolutions of in-commensurability must go beyond mismatches in reference potential.[20]

In fairness, Kitcher does not flatly identify translation with referential agreement. As we shall shortly see, he acknowledges the difficulty and offers a possible way out. Still, Kuhn's general point deserves to be emphasized. His remarks can be turned into an adequacy condition, posing a dilemma-style argument for *any* general account of conceptual change. Purely referential models dodge incommensurability by showing how speakers with radically different theories can nevertheless talk about the same stuff. But reducing translation to referential agreement, Kuhn notes, is problematic. Successful communication requires convergence on what is said about the referents. This requires positing something akin to Fregean senses. But, first, senses are ontologically spooky. Second, and more important, senses are what trigger the problem of incommensurability in the first place, when coupled with the modest holistic assumption that the meaning of theoretical terms is deter-mined by their position within the structure of the entire theory.

In short, here is the dilemma. On the one hand, reference alone is insuf-ficient to characterize inter-paradigm communication and, hence, the ad-vancement of science. On the other hand, richer, intensional accounts solve the problem of communication but lead us back to forms of incommensura-bility. Can we find a conception of progress that is substantial enough to do the trick, but sufficiently slender to avoid unpalatable consequences?

Section 9.6 will take a look at Kuhn's own attempt at taking a stab at the problem—an admittedly metaphorical characterization of the intensional component of translation in terms of culture and the structure of language. Meanwhile, let us focus on Kitcher's proposal that, while not entirely explicit on this score, offers a promising starting point for a solution.

At various points, Kitcher suggests that his notion of reference potential captures something of Frege's non-referential dimension of meaning and the quasi-holistic dictum that concept formation and theory formation go hand in hand.[21] Allow me to elaborate. The notion of sense posited by Frege is supposed to play (at least) two fundamental roles. First, the sense is what

---

[20] My reading of Kuhn's (1982) response to Kitcher is inspired by Carey (1991, 2009). "The problem with Kitcher's argument is that it identifies communication with agreement on the referents of terms. But communication requires more than agreement on referents; it requires agreement on what is said about the referents. The problem of incommensurability goes beyond mismatch of referential poten-tial" (Carey 1991, p. 462).

[21] Contrary to common wisdom, modest holism—typically associated with Quine and Kuhn—was accepted and endorsed by mature logical empiricists such as Hempel (1966).

is "grasped" by anyone who understands an expression; it is the meaning of the term(s). Second, sense encompasses the way in which the reference is expressed; it is the "mode of presentation" of the term(s). Reference potential, Kitcher argues, is akin to sense in this second connotation: it constitutes the mode of presentation of a theoretical term. In addition, since reference potential partly depends on the presuppositions of a paradigm, it becomes clear why scientific terms must holistically absorb theoretical hypotheses. In short, reference potential is the key to understanding conceptual change.

But *what* allows reference potential to fulfill this role? *How* does reference potential specify the mode of presentation of a term? Kitcher's answer appeals to the intentions of speakers and their dispositions to behave:

> Conceptual change in science is to be understood as change in reference potential. The reference potential of a term for a speaker is the set of events which a speaker is disposed to admit as initiating events for tokens of that term. A linguistic community, with respect to a term, is a set of individuals disposed to admit the same initiating events for tokens of the term. An event is the initiating event for a token if the hypothesis that the speaker referred to the entity singled out in that event provides the best explanation for her saying what she did. Explanations are judged by their ability to provide a picture of the speaker's intentions which fits with her environment and history and with the general constraint of the Principle of Humanity. Three kinds of intentions are prominent: the intention to conform to the usage of others, the intention to conform to natural kinds, and the intention to refer to what can be specified. (Kitcher 1982, p. 347)

Plausible as this sounds, intentions, dispositions, and other psychological states are elusive notions. Identifying, characterizing, and attributing these features across agents is a notoriously thorny task. Can we make this any more precise? Sure we can! The remainder of this section shows how black boxes capture reference potentials, explaining how they can play the role of sense *qua* mode of presentation in conceptual change and progress. For now, I focus on phlogiston. Different examples are developed in section 9.5.

Phlogiston has long been a poster child for conceptual change.[22] As we saw in section 4.4 (Chapter 4) and section 9.3, phlogiston theorists purported to account for a number of chemical reactions and, in particular, the process of

---

[22] My condensed presentation is largely based on Kitcher's (1978) insightful overview.

combustion. The core assumption was that all flammable substances are rich in a "principle"—a substance, namely, phlogiston—which is imparted to the air during combustion.

The challenge confronting historians and philosophers of science is to explain how it is that many of Priestley's insights constitute accurate and significant scientific discoveries, given the non-referential nature of its core concepts: phlogiston and derivative notions. This calls for elucidation.

How do we explain the transformations involving, say, the burning of fuel and the heating of metal? Priestley and his colleagues had an interesting story to tell. When we burn a log *en plein air*, the phlogiston contained in the cellulose is released in the atmosphere, leaving ash behind. Similarly, when an iron bar is heated, the combination of heat, metal, and air causes iron to release phlogiston into the surroundings, leaving the "calx" of the metal as a residue. Even more impressively, the theory also provided an account for the reverse of these reactions. Heating the red calx of mercury in an air-rich container results in a combination of mercury and a different kind of air, which Priestley called "dephlogisticated air." How does this work? Phlogiston theory provided an ingenious explanation. Heating the red calx of mercury causes the calx to take up the phlogiston contained in the atmosphere, turning "normal" air into "dephlogisticated air." Priestley's explanations were backed up by successful predictions and retrodictions. For instance, because phlogiston is released during the process of combustion, the residue (ash, calx, or the like) should weigh less than the original substance: wood, metal, etc. For similar reasons, the surrounding air should be altered by the reaction. Both predictions are actually borne out and were experimentally verified.

Regardless of these partial—albeit remarkable—successes, phlogiston theory did not withstand scrutiny. Eventually, the new quantitative chemistry of Lavoisier provided better explanations of these and related phenomena. We now have conclusive evidence against the existence of any principle or substance being emitted in all reactions of combustion. Oversimplifying a bit, from the modern perspective, heating an iron bar produces metal oxide and releases into the atmosphere air that is poor in oxygen. Similarly, heating mercury oxide, what Priestley calls the "red calx of mercury," produces mercury and releases pure oxygen into the surrounding air. In short, phlogiston theory was eventually superseded and replaced by atomic chemistry. But the question remains: how can we explain the partial successes of phlogiston theory, given the non-referential nature of its core concepts?

Kitcher wisely abandons the search for an overly demanding context-independent theory of reference spelling out *the* one reference of "phlogiston" and related notions. He favors a context-sensitive approach, which allows the same expression to vary in reference on various occasions. But when does "phlogiston" refer to oxygen, or to a different natural kind? When does it refer to a nonexistent principle? Sure, intentions have an important role to play. But how do intentions change the functional role of a term within a causal explanation? Let me sketch an answer to this question.

The first key observation is that "phlogiston" acts as a placeholder in a causal explanation represented in a model. To see this, consider the situation from the standpoint of our three-step strategy. First, the inquiries subsumed under phlogiston theory are far from unqualified. The experimental work of Priestley and Cavendish is embedded in a rich thicket of concepts, presuppositions, expectations, and goals. It is only relative to this framework that we can determine and assess the proper explananda, explanantia, and standards for explanation of phlogiston theory. In a word, these requests for explanation are heavily framed. Second, relative to this framework, we have a set of well-formed causal hypotheses that identify the crucial difference-making factors. Third, these causal explanations are represented in a model, which guides the appropriate amount of abstraction and idealization for the inquiry at hand. It also determines which background assumptions may be left implicit and which must be stated explicitly. In short, for all intents and purposes, phlogiston can be constructed as a black box.

It follows from the previous considerations that the term "phlogiston" may figure in two different ways in the context of an explanation. On the one hand, it may designate a mechanism that produces a pattern of behaviors. As such, it is a difference-maker: it is a placeholder that stands in for whatever entity, substance, or process plays the required causal role in sustaining the outcome. Thus, when Priestley maintains that heating the red calx of mercury produces "dephlogisticated air," there must be something that accounts for the observable transformation at hand. Here, "phlogiston" refers to the underlying reaction and, consequently, "dephlogisticated air" corresponds to the substance produced, namely, oxygen. On the other hand, "phlogiston" can also be used as an explanandum, as a description of process to be accounted for, a chemical reaction in need of explanation. For instance, one may ask: how is phlogiston extracted from the log and released into the air? Here, again, "phlogiston" is a placeholder. Yet, it does not stand in for the mechanism producing the reaction. It takes the place of a pattern of behaviors that calls for an explanation. In this latter case, phlogiston is a frame.

Distinguishing between frames and difference-makers helps reveal the complex reference potential of theory-laden terms. When phlogiston is used as a difference maker, it typically picks out whatever produces the behavior identified by the corresponding frame. It may stand in for oxygen, hydrogen, or other substances unknown to Priestley and other phlogiston theorists. It may also stand in for nothing at all, when the frame specifies an explanandum that is not produced by anything. For instance, one may ask: in what ways does phlogiston interact with the cellulose in the wooden log, such that it gets released into the surrounding air? Here phlogiston is a non-referential term. It picks out nothing at all because there is nothing at all to be explained. In contrast, when used as a frame, "phlogiston" may stand in for the combustion-events or for other chemical reactions in need of explanation. Some frames provide well-formed explananda, from a contemporary perspective. What principle or substance is emitted in all cases of combustion? Answer: none. Other explananda can be dismissed as ill-posed, from a modern standpoint. What is the atomic number of phlogiston? How does phlogiston interact with hydrogen? Answer: these questions have no clear place in our contemporary chemical theories.

In sum, Kitcher is correct that conceptual change should be understood as change in reference potential, that reference potential captures Frege's notion of sense *qua* mode of presentation, and that speakers' intentions, as well as other psychological notions, play a significant part in the determination of reference. But reference potential itself can be further broken down in terms of its functional role in the context of an explanation. Identifying theoretical terms with black boxes distinguishes their act as frame and difference-makers. The three steps of the black-boxing strategy—framing, difference-making, and representation—help us see when reference remains constant across paradigms ("planet"), when reference potential needs to be reassessed in the light of future evidence ("atom"), and when the term is better dropped altogether ("phlogiston"). In short, "Kuhnian" conceptual change depends on how a black box is packed or unpacked within its framework.

This section showed that the black-boxing strategy provides the resources to elaborate the notion of reference potential, refining our understanding of non-cumulative scientific progress. Intentions are elusive notions. Black boxes help make them more perspicuous. At the same time, the notion of reference potential sheds light on black boxes. The following section applies the conceptual resources inherited from our discussion of conceptual change to re-examine the role of black boxes in the success stories of Darwin and Mendel, the mistakes of behaviorism, and tales like neoclassical economics, where the final verdict is yet to be uttered.

§9.5. Black Boxes and Reference Potential, Part II

Chapter 3 presented four episodes in the history of science: Darwin and Mendel's momentous contributions to biology, Skinner's behavioristic psychology, and Friedman's "as if" economic models. Chapter 6 revised these case studies in the light of our three-step black-boxing recipe. I concluded that the success and limitations of black boxes can be explained in terms of framing, difference-making, and representation. Our excursus into theory change, conceptual refinement, and scientific progress provides the resources to develop our analysis by addressing an important question: how should we understand the relation between black boxes across paradigm shifts?

§9.5.1. Gemmules, Variation, and All of That Darwinian Jazz

Let us begin with evolutionary biology. The relation between Darwin's own approach and contemporary evolutionary theory is not ordinarily viewed as one of "incommensurability." Indeed, whether Kuhn's account of scientific revolutions applies to biology at all can be, and has been, questioned.[23] Still, as we shall presently see, the unfolding of the theory of evolution since its inception raises puzzles similar to the ones discussed in previous sections.

Consider Darwin's own views of inheritance, as presented in Chapter 27 of *Variation of Plants and Animals under Domestication* (cf. §3.2). According to his "provisional hypothesis of pangenesis," both the transmission of heritable qualities and the unfolding of ontogeny are caused by invisible particles. These microscopic entities, called "gemmules," are thrown off by cells and, when supplied with sufficient nutrients, they can multiply by self-division. Just a few years after the publication of this conjecture, Weismann

---

[23] For a clear and insightful discussion of this issue, see Godfrey-Smith (2003).

conclusively showed that gemmules do not exist. In this respect, they resemble phlogiston. Just as phlogiston is the heart of Priestley's theory of combustion, gemmules play a central role in Darwin's views on inheritance. As we saw, Priestley's theory raises a puzzle: how can a non-referential concept—namely, phlogiston—contribute to scientific discoveries? The same question can be asked with respect to Darwin's pangenesis, with the additional complication that, contrary to phlogiston theory, Darwinian evolution has not been superseded. What is going on here?

One option is to flatly ignore Darwin's claims in *Variation* and focus on the earlier, more influential framework in *Origin*, where the English scientist remained agnostic on the nature of the mechanisms of inheritance. Rather than solving the conundrum, this whiggish historical reconstruction merely sweeps the dust under the rug. After all, Darwin did make this mistaken claim about the nature of inheritance. And his errors did not affect the explanatory success of his theory. The central question is: *why* did these mistakes leave the overall success of the theory unscathed?

A much more promising strategy borrows Kitcher's context-sensitive notion of reference potential. From this standpoint, we can recognize how the reference of various tokens of "gemmules" varies with the circumstances. Thus, for example, when Darwin speculates that "the gemmules in their dormant state have a mutual affinity for each other, leading to their aggregation either into buds or into the sexual elements" (*Variation*, p. 374), it seems reasonable to interpret him as asserting a mistaken claim about nonexistent entities, namely, gemmules. Now, contrast this with Darwin's assertion that gemmules "are supposed to be transmitted from the parents to the offspring, and are generally developed in the generation which immediately succeeds, but are often transmitted in a dormant state during many generations and are then developed" (*Variation*, p. 374). By applying the principle of humanity, we may surmise that, here, "gemmule" refers to (what are now known as) genes and other cytological gears, making Darwin's statement true, or approximately so. Following Kitcher, we conclude that this referential heterogeneity makes the concept of gemmule "theory laden."

The black-boxing strategy extends and develops this point, further explaining why and how reference potential works the way it does. Specifically, when does "gemmule" refer to genes? When does it fail to refer? The answer to this question should now be evident. Darwin's notion of gemmule is part of a much broader theoretical framework. His explanandum is not a generic: how are traits inherited? As we saw in section 6.2 of Chapter 6, his objective is a particular account of inheritance, given certain known facts about physiology, given that some of these traits must make a difference to fitness, given observed distributions within and across species, etc. Against this backdrop, it becomes clear how, in some cases, the entities responsible for these patterns are explananda. What is the structure of the mechanisms responsible for the transmission of trait *t* across generations? In *Origin*, Darwin candidly admits that he has no answer to offer. In *Variation*, he sketches a speculative proposal, which turned out to miss the mark. In neither case does it seem plausible to claim that Darwin has genes in mind. In other cases, however, gemmules are used as explanantia, as when Darwin appeals to these invisible particles to explain the intra-generational transmission of traits. Here, by applying the principle of humanity, we can take "gemmule" to stand in for whatever mechanism(s) fulfill the appropriate causal role: genes and other cytological gears. In the first situation, gemmule is a frame. In the second, it is a difference-maker. In both cases, it is a placeholder in a causal explanation represented in a model. Either way, it is a black box.

These considerations are perfectly general and can be applied beyond gemmules. Evolution by natural selection is a balanced blend of various ingredients: competition, variation, fitness, heritability, and much else. All of these components can be broken down further into more fundamental constituents. From the privileged vantage point of contemporary evolutionism, Darwin's attempts to do so obtained mixed results. At times, he was remarkably successful, as when he distinguished between variations that affect fitness and variations that do not. Other endeavors were more problematic, as we have seen with his attempts to unravel inheritance. Darwin at first lacked a solution, and his subsequent proposal missed the mark. Nevertheless, by viewing Darwin's constructs as black boxes, as frames and difference-makers represented in models, we can see how his theory is a mix of true claims about genes and cytological gears, retained and developed in subsequent elaborations, and false claims about pangenesis, soon to be discarded.

§9.5.2. Genes: Mendelian, Molecular, or Manifold?

Genes provide another interesting case study for conceptual change. In the course of the twentieth century, the reference potential of the term "gene" has been significantly altered in response to experimental and theoretical innovations. This raises interesting puzzles. On the one hand, geneticists since Mendel have always been talking about the same things, namely, chromosomal segments. On the other hand, it is clear that, over decades of research, tokens of "gene" refer to distinct entities. On various occasions, different chromosomal segments may be picked out. The question becomes: what are the principles that determine the specific referent of each token?

In his detailed analysis, Kitcher (1982) distinguishes two biological characterizations of "genes." The first strategy, prominent in Mendelian genetics, identifies genes by focusing on their function in producing phenotypic effects. In its early usage among Morgan's *Drosophila* group, "gene" or "factor" referred to a set of chromosomal segments, each of which plays a distal causal role in the determination of phenotypic traits. Because of ambiguity in the specification of these functional roles, the concept *gene* rapidly acquired a heterogeneous reference potential, where different tokens could pick out different segments, in a hypothesis-relative fashion. This classical gene concept was molded into its definitive form in the 1950s, when Benzer introduced "cistrons" into the picture. The second approach, common in molecular biology, identifies genes by focusing on their proximate action. On this view—first articulated in the 1930s in the context of Beadle's "one gene–one enzyme hypothesis"—chromosomal segments are pinpointed according to their functional roles at the earlier stages of the process, as opposed to using relatively indirect mutation and recombination tests.

These well-known considerations, Kitcher argues, show the existence of many concepts of gene, determined by alternative decisions at the phenotypic level. These concepts are not in competition. Different ways of dividing chromosomes—on a spectrum ranging from codons coding for single amino acids to lengthy DNA sequences coding for multi-polypeptides—will be best suited to serve one's interests depending on the particular research project at hand. From this standpoint, asking which criterion of segmentation corresponds to *the* concept of gene is an ill-posed question. As anticipated in section 9.3, the term is theory-laden and its reference varies across contexts.

The black-boxing strategy preserves the spirit of Kitcher's historico-philosophical reconstruction. In addition, it further explains how and why the notion of reference potential captures Frege's sense *qua* mode of presentation without lapsing into the quagmire of radical incommensurability.

Alternative gene concepts underlie different theoretical and experimental purposes. Classical genetics and molecular biology are not addressing the same questions.[24] This becomes crystal-clear at the first framing stage of the black-boxing recipe. Classical geneticists were perfectly aware that genes are not sufficient causal bases for complex phenotypic traits such as eye color or wing shape. From a Mendelian perspective, genes are supposed to account for variation among members of a population. Molecular genetics, in contrast, presupposes an altogether different framework, which aims at unraveling the mechanisms underlying ontogeny. With this in mind, it is hardly surprising that genes do not play the same difference-making role in these causal explanations. Classical genetic models represent genes as difference-makers for phenotypic traits. Molecular biological models, in contrast, represent genes as difference-makers for the production of polypeptides. Hence, these two accounts should not be viewed as competing. Classical and molecular genetics should be assessed independently, on their own ground.

At this point, some attentive readers may have noted, we face a problem of a different sort. If classical and molecular biology presuppose different concepts of gene, then are they really talking about the same kind of stuff? Following Putnam's insight, developed in section 9.3, it may be tempting to note that both theories refer to the same sets of entities, namely, genes. But is this a plausible response? As Kitcher rightly notes, asking which concept of segmentation corresponds to "the" concept of gene is an ill-posed question. But, then, what does it mean to say that both theories are talking about genes? What kind of genes? The issue, simply put, is that we are now led back into the swamp of meaning incommensurability. If the meaning of theoretical terms is theory-dependent, the meaning will change with the theory. But this seems simply preposterous. As many authors have noted, molecular biology extends and deepens the explanations of classical genetics. But this presupposes that there is a common measure to assess them. What is it? Once again, black boxes to the rescue.

Both classical genetics and molecular biology treat genes as black boxes. Importantly, they are not the same black box. Morgan and colleagues frame and represent genes as the units of mutation, recombination, and phenotypic function. Within the framework pioneered by Beadle and later refined by Watson, Crick, and subsequent molecular biologists, genes are framed and represented as functional units responsible for transcribing protein. At the same time, at broader levels of description, viewed as the mechanisms

---

[24] This point, presented cogently albeit in a different context, by Waters (1990, 2007), will be developed, in greater detail, in Chapter 10.

underlying inheritance and variation, all these units can be subsumed under a common description. It is these broader frames and difference-makers that constitute the *trait d'union* between theories that make very different assumptions. What is the relation between these levels of description? This is a question that will be addressed in Chapter 10. And, once again, the notion of incommensurability will play a central role. Before getting there, however, we still have a few more examples to cover.

§9.5.3. Where Did Behaviorism Go Wrong?

Black boxes capture conceptual progress in Darwin and Mendel. Can analogous considerations shed light onto the strength and weaknesses of sophisticated radical behaviorism? Skinner is frequently faulted for his decision to black-box mental states. As we saw in section 6.4 of Chapter 6, this is a mistake. Skinner did deliberately introduce abstractions and idealizations in his psychological models. His admittedly oversimplified characterization of stimuli and behavior is a good illustration. But this, in and of itself, is no more problematic than, say, Darwin's oft casual appeals to traits, fitness, and the mechanisms of inheritance. Furthermore, unlike Darwin, Skinner's explanations do not involve any systematic introduction of nonexistent entities or non-referring terms. This basic parallel raises a third problematic observation. The principle of humanity permits a charitable reinterpretation of some tokens of "phlogiston" and "gemmules" to refer to oxygen and genes, respectively. Could one not rephrase Skinner's behavioral states so as to include psychological and neural processes, as we understand them today? The problem is that now it seems hard to celebrate Darwinism as a success story while discarding behaviorism as an outdated theory of mind. But this is blatantly absurd! Where did our analysis go wrong?

The appropriate reaction is not to throw in the towel and settle for an "anything goes" relativism. Skinner was, indeed, guilty of mistakes that did not undermine Darwin's and Mendel's approaches. But, in order to pinpoint these shortcomings, we need to move beyond referential models of meaning, even when reference is—rightly—understood in a context-sensitive fashion. The lack of progress that eventually turned radical behaviorism into a regressive research program involves not what the theory talks about. All its key terms are perfectly referential. The problem pertains to the structure of Skinner's causal explanations of human conduct. Allow me to elaborate.

Sophisticated behaviorism can be portrayed as an attempt to account for the conduct of intelligent beings without appealing to irreducible mental phenomena. Recall from section 6.4 the typical causal explanation in psychology—$E \to M \to B$—constituted of environmental events (E), mental events (M), and behavioral events (B). Skinner's theoretician's dilemma purported to show that the intermediate link M could be eliminated without any substantial loss of explanatory power. This involves showing that mental states are dispensable *qua* explanantia and *qua* explananda, since, in our chain, they are explained by the environment and they explain behavior. We can now state clearly why both claims are fundamentally problematic. Skinner is correct that environmental stimuli play a decisive role in the production of mentality. The cool breeze is partly responsible for my desire to wear a sweater. His mistake was overlooking the irreplaceable role of logical, cognitive, and linguistic competence, which are required to frame and causally explain action and other forms of human conduct. All the right ingredients are in place. Yet, the explanandum is poorly framed and the choice of difference-makers is inadequate. This shows that scientific progress goes well beyond successful reference. The causal representation must also be adequate for the true advancement of science.

In sum, Skinner's abstraction and idealization in psychological explanation was perfectly acceptable; indeed, it was fruitful. Like the main ingredients of Darwin's evolutionary theory and Mendelian genetics, mental and behavioral states are black boxes. Still, these placeholders were not appropriately constructed. The box of mentality lacks cognitive structure, and the box of behavior ignores the framing and difference-making role of psychological states. This is where behaviorism went wrong. Black-boxing is clearly ubiquitous in science. But not all black boxes are made equal.

§9.5.4. Two Concepts of Utility

Finally, consider the black box of neoclassical economics, which provides the most interesting and controversial example because of its open status. As discussed in section 6.5 of Chapter 6, Friedman and Savage offer a unified explanation of economic behavior by positing a single algorithm for the maximization of utility. According to their influential proposal, the difference-maker underlying all economic choices—including both risky and riskless ones—is the agent's assignment of utility to alternatives. Fans of neoclassical economic theory praise the simplicity, elegance, and mathematical rigor of this simple model. Detractors criticize the lack of realism. Rather than positing a highly idealized notion of utility and abstract conditions for rationality, psycho-neural economists contend, we are better off by looking at the mechanisms of behavior that are actually implemented in the human brain.

If we look at the situation from the standpoint of theory change, the relation between neoclassical and psycho-neural economic approaches might appear like a textbook example of incommensurability. Both stances assign a central role to utility. But is it the same construct? The notion of utility introduced in neoclassical economics is an abstract mathematical value with no physical counterpart. Following the groundbreaking insights of Pareto at the dawn of the twentieth century, neoclassical economists view utility as an ordinal ranking revealed by preference, with no cardinal value. Psycho-neural economists, in contrast, identify utility with an actual neural value, instantiated in the brain, and measurable, at least in principle.[25]

These two theoretical constructs could hardly be more different. This raises a quandary. As we saw in our discussion of genetics, asking which chromosomal segments correspond to "genes" is an ill-posed question. Wondering about which definition of utility is the "correct" one fares no better. But then, what are we to make of the heated debates in contemporary economics? The reference of "utility" is clearly different, but the alternative theories nonetheless compete. Once again, this goes to show that commensurability, communication, and progress transcend shared denotation.

Reference potential is a first step toward a solution. The short answer is that, depending on the context, tokens of the term "utility" may refer to various concepts across subfields of economics. "Utility" may pick out a mathematical construct, a neural mechanism, a set of beliefs and desires, or something altogether distinct. This effectively captures how different theories posit a different conception of utility. But this is only half of the story. The other side involves the shared measure, the common denominator that makes these theories competitors. If they are talking different notions of utility, in what sense are they disagreeing? Black-boxing provides a simple and effective solution.

The common denominator between our two economic approaches is not provided by shared reference—as just noted, different uses of "utility" pick out

---

[25] These presuppositions, often left implicit, are discussed explicitly in Glimcher (2011).

different concepts. The theoretical overlap is provided by the structure of the black box. Viewed as a black box, utility may figure in economic explanations both as explanans and explanandum. When we ask what determines my selection of, say, coffee over tea, utility is an explanans, a difference-maker that explains the choice itself. In contrast, when one inquires into the nature of the mechanism underlying choice, utility becomes a frame, it is the object of the explanation, the explanandum. With this distinction in mind, it becomes clear that alternative theories provide competing explanations of utility. Neoclassicists opt for a highly idealized "as if" mathematical approach. Psycho-neural economists spell out a mechanistic model. These are distinct difference-makers, represented in alternative models. At the same time, both theories have a single objective in mind: an explanation of the same economic behavior. The use of utility as an overarching frame to pinpoint the behavior under investigation is the common factor, making the two frameworks different explanantia of the same core explananda. In short, across neoclassical and psycho-neural economics, utility spells out the same frame, but alternative difference-makers.

Time to take stock. Reference potential performs a role analogous to Frege's mode of presentation, making concepts commensurable across paradigm shifts. The black-boxing strategy explains how this can be done in a rigorous fashion, without positing intensional entities, without lapsing into meaning incommensurability, or other unpalatable forms of holism. Gemmules, genes, mental states, and utility are all theory-laden concepts in virtue of their heterogeneous reference potentials. What determines which of these uses is preserved across theories and paradigm shifts is the structure of these black boxes—framing, difference-making, and representation. We can thus make sense of Hempel's dictum that "concept formation and theory formation go hand in hand." Theories presuppose concepts, and concepts require a theoretical framework. Black-boxing shows how both points can be held simultaneously, without getting us stuck in a vicious circle.

§9.6. Incommensurability, Progress, and Black Boxes

In his 1982 article, "Commensurability, Comparability, Communicability," Kuhn asks: what is it that translation must preserve? Not merely reference, he notes, since reference-preserving translations may well be incoherent or incomprehensible. What about requiring that translation preserve sense or intension? While Kuhn, at one time, endorsed this strategy, he now recognizes that this cannot be quite right either. Without denying the importance of addressing the thorny issue of meaning, Kuhn concludes the aforementioned article by trying to avoid semantic talk altogether, replacing it with a quasi-metaphorical discussion of how members of a language community pick out the referents of the term they employ. How can it be, Kuhn wonders, that people with very different criteria regularly pick out the same reference for their terms? A preliminary answer notes that language is adapted to our social and natural world, and the world is such that it does not present objects and situations that would trigger different identifications. But this begs a further, deeper question: what must speakers with disparate reference-determining criteria share for them to count as speakers of the same language, to be members of the same linguistic community?

> By now it must be clear where, in my view, the invariants of translation are to be sought. Unlike two members of the same language community, speakers of mutually translatable languages need not share terms. "Rad" is not "wheel." But the referring expressions of one language must be matchable to coreferential expressions in the other, and the lexical structures employed by speakers of the language must be the same, not only within each language but also from one language to the other. Taxonomy must, in short, be preserved to provide both shared categories and shared relationships between them. Where it is not, translation is impossible, an outcome precisely illustrated by Kitcher's valiant attempt to fit the phlogiston theory to the taxonomy of modern chemistry. (Kuhn 1982, p. 683)

In my opinion, Kuhn is wise in eschewing talk of intentions, and he is correct about the importance of preserving taxonomy. Yet, what exactly does it mean to "preserve taxonomy"? On what grounds can we compare and contrast taxonomies across competing theories and paradigm shifts?

Kitcher's notion of reference potential is a solid starting point in answering this question. He advocates for the importance of examining the intentions of speakers and to interpret them via a "principle of humanity." But intentions are notoriously slippery notions. I have tried to show that black-boxing provides a strategy to refine these intuitions, making them more objective and precise. Preserving taxonomy, from our perspective, is ultimately a matter of identifying the right frame or difference-maker, while setting aside the precise nature of their realizers. Darwin, Mendel, and Crick have very different conceptions of gene in mind—indeed, Darwin and Mendel did not talk about "genes" at all. The same can be said about Priestley and Lavoisier. These authors all share a coarse description of a set of explananda and the underlying mechanisms. In short, Kuhn's preservation of taxonomy can be fruitfully understood as the unfolding of black boxes across theories. These boxes capture the structure that remains constant across paradigm shifts, for instance in the shift from the classical physics of Newton to Einstein's relativistic framework.

Time to tie up some loose ends. An account of progress is a central component of any serious analysis of science. For a long time, philosophers assumed, more or less implicitly, that the advancements of science could be depicted as a gradual accumulation of truth and knowledge. Kuhn, Feyerabend, and their colleagues are rightly credited with a decisive dismissal of this simplistic view. However, their *pars construens* was not quite as persuasive as their *pars destruens*. Referential models, the most popular solution to radical holism and meaning incommensurability, are not structured enough to capture scientific progress. Successful translation requires more than mere agreement on reference. The black-boxing strategy attempts to blend the objectivity of purely referential models with a mild—but inevitable—form of meaning holism to sketch a notion of progress that allows us to compare frameworks and avoid the bogey of incommensurability.

In addition, a discussion of progress sheds further light on historical examples. Darwin's and Mendel's work reveals how, as long as the structure of the black box is identified correctly, failure to pinpoint the underlying mechanisms does not affect the success and fruitfulness of a scientific hypothesis. Skinner's shortcomings warn of the danger of mis-framing and mis-representing a causal explanation. Finally, the debate between classical and psycho-neural economists shows that identity of reference is neither a necessary nor a sufficient condition for progress. Utility is understood very differently across these frameworks. But, at a coarse level of description, it plays the same functional role as frame and difference-maker.

One final remark. Hopefully, the discussion in the previous three chapters shows that my analysis of black boxes is not an attempt to rebuild philosophy of science entirely from scratch. This prominent albeit neglected construct captures a number of traditional themes—such as mechanisms, emergence, and progress—and recasts them in a new light. At the same time, talk about meaning, reference, and incommensurability does not necessarily drag us back to hackneyed debates in the philosophy of language. Be that as it may, we can now move on to the final leg of our long journey together. True scientific achievement is a combination of decomposition, as emphasized by reductionists, and autonomy, a trademark of antireductionism. It is time to apply all of our insights to this central debate in current philosophy of science.

10

Sailing through the Strait

Ich habe dir nicht nachgestellt
Bist du doch selbst ins Garn gegangen,
Der Teufel halte, wer ihn hält!
Er wird ihn nicht so bald zum zweiten Male fangen.

—J.W. Goethe, Faust, 1426–1429*

§10.1. Back to Bricks and Boxes

Our long excursus into the nature and structure of black boxes began with an old dusty image, which was mainstream well into the twentieth century and is still popular in some circles. This is the figurative depiction of science as a slow, painstaking accumulation of truths. The goal of scientific research, from this hallowed perspective, is to provide an accurate and complete description of the universe, or some portion of it. In this sense, the development of science is akin to the erection of a wall or the tiling of a mosaic. The building blocks of science are basic facts about the universe we inhabit.

Over the years, scholars—scientists, philosophers, historians, and many others—have vocally denounced the misleading nature of this analogy. At best, it is a drastic oversimplification. At worst, it utterly misses the mark. Either way, it should not be taken too seriously. Popper's suggestive image of the scientist as a knight in shining armor battling the evil forces of darkness has followed suit and faded away. Of course, truth, knowledge, and objectivity remain important goals for the scientific enterprise. But they are only one side of the story. The remainder involves what we do not know, what we cannot grasp, what we get wrong. In a word, what is missing from the "brick-by-brick" model of science is the productive role of ignorance.

---

* "You were not caught by my device // When you were snared like this tonight. // Who holds the Devil hold him tight! // He can't expect to catch him twice." Translation by W. Kaufmann.

Black Boxes. Marco J. Nathan, Oxford University Press. © Oxford University Press 2021.
DOI: 10.1093/oso/9780190095482.003.0010

The cumulative image remains popular among the general public. Journalists, politicians, entrepreneurs, and other non-specialists typically presuppose, more or less consciously, the vision of science as a gradual accumulation of truths. The source of this misconception, I suggested at the outset, is the standard packaging of the inquiry. Textbooks and lectures introduce science as a bunch of laws, theories, facts, and other cut-and-dried notions to be memorized and, eventually, internalized uncritically. This resilient stereotype trickles down from schools and universities to television shows, newspapers, magazines, and other channels that perpetuate it among the public. Only the small portion of students who attend graduate school and pursue research careers is exposed to the true face of science. Real practice at the bench is quite different, and way more interesting, than the ossified caricatures presented to outsiders in journals, books, and conferences.

At the same time, the textbooks from which students learn science are written by those same specialists who, when pressed on the issue, adamantly reject the brick-by-brick conception. This raises some questions. Why are experts perpetuating an image of science that they themselves disavow? Are there more fruitful ways of exposing the new generations to the fascinating world of science? Can we do a better job at popularizing our research?

My answer can be broken down into a *pars construens* and a *pars destruens*. The negative claim presents the two dominant models of science—reductionism and antireductionism—as ultimately unsatisfactory. This is because, at a general level of description, neither resolves the tension between two intuitive aspects of scientific practice. Reductionism captures the insight that breaking a complex system down to a more fundamental level always provides valuable information. But this seems to leave no room for productive ignorance, that is, the autonomy of higher-level explanations. Sure, from a reductionist perspective, ignorance may be useful when it points out what still needs to be done—in this sense, it may not be a hindrance. Still, knowing is always better than not knowing, and replacing ignorance with more details invariably constitutes a step forward. Antireductionism, on the other hand, legitimizes the role of ignorance and autonomy in science, but has trouble explaining how macro-descriptions can be objectively superior to their micro-counterparts. When we focus on more local debates, modest reductionism and sophisticated antireductionism substantially overlap, making the dispute largely semantic. After decades of debate, philosophy of science appears to be stuck between Scylla and Charybdis. Can we find a different route forward?

My positive proposal is simple. Over the centuries, science has developed a well-oiled technique for turning ignorance and failure into knowledge and success. This practice hinges on the construction of black boxes.

References to black boxes are ubiquitous across the sciences and the humanities. Intuitive familiarity with the expression may well be the reason that compelled you to pick up this book in the first place. And the general idea should be clear and intuitive enough. Black-boxing a phenomenon involves isolating some of its core features in a way that it can be assumed without further explanation, justification, or detailed structural description.

This, in a nutshell, is a black box. But that cannot be the entire story. Even a cursory glance at the history of science makes it clear that not all black boxes are the same. Some work perfectly well; others muddy the waters. Some are constructed out of necessity; others are the product of error, laziness, or oversight. Some unify; others divide. In short, the well-known tales of Darwin, Mendel, Skinner, and Friedman reveal that black-boxing is more complex, nuanced, and powerful than is typically assumed.

At root, black boxes are placeholders. Yet, the role of placeholders in science is more prominent, interesting, and subtle than people realize. For one, there are (at least) two kinds of placeholders. *Frames* stand in for behaviors in need of explanation. *Difference-makers* stand in for whatever produces the behaviors in question. With this distinction in mind, I broke down the construction of a black box into three constitutive stages: framing, difference-making, and representation. This led to the definition of a black box as a placeholder in a causal explanation represented in a model. The simplicity of this one-liner is deceptive: clarifying the steps and underlying presuppositions required long and technical discussions.

The second part of the book put this definition to use. First, I reassessed our case histories in evolution, genetics, psychology, and economics from the standpoint of black boxes. This motivated my contention that the history of science is a history of black boxes. Next, I argued that black-boxing constitutes a "diet" form of mechanism. Finally, I applied black boxes to two substantial philosophical issues: emergence and scientific progress.

It is time to wrap up our discussion by cashing out my last two promissory notes. At the end of Chapter 1, I asked two related families of questions. First, can we avoid both Scylla and Charybdis? Is there an alternative explanation of the nature and advancements of science? What is it? Second, how does science turn ignorance into knowledge? Black boxes are the key to answering both sets of inquiries. The goal of this chapter is to substantiate these claims.

How does black-boxing reframe reductionism and antireductionism? How does it capture the role of ignorance in science?

Here is the plan. Section 10.2 recaps what got us into the current stalemate. Section 10.3 argues that black-boxing captures the advantages of both reductionism and antireductionism, while eschewing their most troubling implications. Section 10.4 elucidates my claims by illustrating them with our historical examples. I conclude that, as I mentioned at various stages throughout the book, the history of science really is a history of black boxes. Section 10.5 wraps up the discussion by cashing out the notion of productive ignorance that is missing from our extant models. By that point we will be ready for farewells.

§10.2. The Quagmire

Chapter 2 portrayed contemporary philosophy of science as figuratively sailing between two hazards: the Scylla of reductionism and the Charybdis of antireductionism. Almost sixty years after the publication of Nagel's *The Structure of Science*—the epitome of the positivist ideal—the battle between friends and foes of reductionism rages on. No winners or losers are in sight, but both armies look fairly worn out. Before exploring alternatives, it might be worth recalling what got us into this quagmire in the first place.

Positivism maintained that all of science can be reduced to fundamental physics, at least in principle. Reduction was conceived as a deductive relation between theories, characterized as interpreted axiomatic systems.

This classical model of reduction has fallen on hard times, eroded by powerful objections: the multiple-realizability of many prominent higher-level kinds; the lack of appropriate laws, generalizations, and bridge principles; and the shortcoming of "syntactic" conceptions of theories. Taken together, they mount an overwhelming case against Nagel's derivational model of reduction. The demise of classical reductionism divided philosophers into opposite camps, depending on their methodological inclinations.

One strategy aims at refining the reductive framework. The story goes something like this. The classical model had the right ideal in mind, but failed to implement it properly. Reduction should not be cashed out as a logical relation between interpreted axiomatic systems. The main issue is a question about *explanation*. Lower-level descriptions, neo-reductionists maintain, invariably deepen and enrich our understanding of systems, technically rendering higher-level depictions disposable. Thus, all of science can, in principle, be reduced to physics. But this is not because theories in the special sciences can be logically deduced from physical laws, or other underlying generalizations. They cannot. The reason is that, barring pragmatic considerations of computational tractability and convenience, micro-descriptions provide the best explanations: the deeper, more powerful, more comprehensive models. In its most radical, ambitious formulations, physics regains its role as a true "theory of everything," the paradigm and limiting case of all of science, just like it has been for much of its history.

The second, alternative strategy follows a diametrically opposite path, eschewing reductionism *tout court* and treating the special sciences as "autonomous." More specifically, antireductionists agree with neo-reductionists on the shortcomings of the classical model. However, they diverge on the diagnosis and its implications. It is false, the classic antireductionist argument runs, that higher-level explanations are invariably empowered when they are reassessed from a lower-level standpoint. Sure, in many cases, structural details are crucial for addressing some macro-explanandum. But this downward trajectory does not extend all the way down to the foundations of the discipline itself, let alone to physics. There is a threshold past which adding structural information does not enhance higher-level explananda. If anything, unnecessary details end up hindering connections and muddying the waters. This, in a nutshell, is the *autonomy* of the special sciences.

Before moving on, it is useful to clarify, one final time, the nature of the debate. What is at stake in the divide between friends and foes of reduction?

First and foremost, the contention is independent of the truth or falsity of physicalism. Sure, a handful of philosophers still resist the reduction of mental to physical states on the basis of Cartesian or Neo-Cartesian dualistic assumptions. But most discussants, reductionists and antireductionists alike, tend to be thoroughgoing materialists. The physical supervenience of biology, neuropsychology, economics, and other special sciences is typically taken for granted. The crux of the matter is what exactly follows from this assumption: should materialism be "reductive" or "non-reductive"?

Second, the fruitfulness of decomposing complex systems into more basic constituents is not in question. This is not a contentious methodological strategy.[1] As new mechanists have helped clarify, it is a truism that everyone, no matter their ontological inclinations, should take for granted.

As I said, the heart of the contention concerns explanation. Do micro-details invariably deepen macro-explananda? Are higher-level descriptions autonomous from lower levels? Are physical accounts of Putnam's square peg deeper or overkill? Is his geometrical account incomplete, or does it strike just the right balance? Is it true that all molecular findings enhance classical genetics, or do we reach a point where further additions become unnecessary? These are the questions that have divided philosophers for decades.

So, who wins, the reductionist or the antireductionist? Is it going to be the rock shoal or the whirlpool? To begin, note that pros and cons of both stances, presented at the most general level, are unsurprisingly specular: one is the mirror image of the other.

Reductionism provides a clear-cut account of how discoveries at the lower levels may, in principle, inform and extend higher-level theories. Suppose that one is trying to explain why table salt dissolves in tap water (cf. §2.3, Chapter 2). Intuitively, the simplest strategy is to look at the chemical composition of these elements. But why should this be so? Why would an analysis of H_2O and $NaCl$ shed light on the behavior of salt and water? Reductionism has a prima facie compelling story to tell. If salt and water are reducible, in some relevant sense of the term, to their molecular structure, then there is no mystery why the nomological connection between H_2O and $NaCl$ explains their respective behaviors. The apparent simplicity of this illustration should not mislead us. Similar reasoning also sheds light on how biochemistry informs functional biology or how neuroscience contributes to psychology.[2] This is what makes reductionism such an intuitive stance.

Antireductionism, in turn, is fueled by an equally compelling observation: the autonomy of higher-level models. Consider, once again, the example just rehearsed. The relation between salt and water is correctly described as one of *solubility*. This characterization appears to be perfectly adequate and exhaustive. Furthermore, increased accuracy does not come for free. Solubility is a relatively simple property. But it has the virtue of being quite general. It captures the relation between salt and water, but it is also applicable

---

[1] Thus, for instance, Mayr (2004), an outspoken antireductionist, calls this basic tenet "analysis" and contrasts it explicitly with "reduction."

[2] I explore these issues, more systematically, in a series of articles: Nathan (2012, 2015b); Nathan and Del Pinal (2016, 2017).

to sugar and tea, stevia and soda, and many more substances. Details about H_2O and $NaCl$ increase the depth of the explanation. But adding precision will *ipso facto* impact its breadth, its generality. Given that salt and sugar have different chemical structures, making the model too similar to the former substance will make it inapplicable to the latter. In short, there is a trade-off between accuracy and generality, and the golden mean is not invariably, or even typically, found at the lowest, most fundamental levels. This, simply put, fuels the methodological autonomy of the special sciences, the heart and soul of contemporary antireductionism.

Viewed in this light, the choice between reductionism and antireductionism may well turn out to be an ill-posed one. Sophisticated versions of both these influential stances stress different, equally legitimate aspects. Philosophy tends to view them as incompatible, but this is a mistake. As we saw, when we focus on concrete examples, initially apparent differences fade away. Try asking whether current science better conforms to reductionist or antireductionist standards, and the matter tends to lose substance.

The stalemate becomes especially evident in the biological sciences, where much is known about the structural implementation of function. Can all biology be reduced to molecular biochemistry? The trouble with this question is that it requires and presupposes a rigorous definition of "molecular property" that we currently lack and will likely never be found. The crux of the disagreement is whether our best biological models are molecular, and this ultimately depends on whether we classify explanations as "molecular." This is a semantic, terminological issue, not a substantive one.

Philosophers of mind, psychology, neuroscience, and economics should accept the dire lesson from their colleagues working on biology. Empirical discoveries are unlikely to solve long-standing philosophical disputes over the mind-body problem or the relation between economics and psychology. The reason is not the—indisputable—complexity of these fields. The issue is the nature of reduction which, contrary to common wisdom, is murky. All of these observations point to the same conclusion: it is time to move on.

§10.3. Rocking the Rock, Weathering the Whirlpool

Section 10.2 recapped the argument, originally spelled out in section 2.6 of Chapter 2, that the current standoff between reductionism and antireductionism is not as substantive as it is typically taken to be. Our present goal is to blaze an unbeaten trail, find an alternative way forward. My suggestion is to reframe the current debate from a different perspective: the standpoint of black boxes.

Traditionally, philosophers pitch autonomy and reduction against each other. Following Putnam's lead, autonomy is conventionally defined as the rejection of reduction. Reduction thus *ipso facto* becomes the denial of autonomy. As just presented, these two stances are obviously mutually incompatible, in virtue of their meaning alone. But could this be a mistake? Why not accept both tenets simultaneously? Could we not maintain that descending to a lower explanatory level invariably enhances explanatory power, without thereby giving up on the autonomy of higher-level description? Can we not have our cake and eat it too? I want to argue that, yes, we can.

How should the issue of (anti)reduction be repackaged so as to make it more substantive and significant? This section sketches an answer to this question, based on the black-boxing recipe developed in previous chapters. The following section (§10.4) will illustrate the proposal by applying it to the case studies that accompanied us during our journey together.

Let us consider the relation between higher and lower levels—intra-field or inter-field, it does not matter.[3] On the one hand, it is clear that micro-depictions provide details that deepen macro-explanations in various ways. This is hardly a shock: discovering the workings and implementation of a complex system is key to understanding its behavior. This is the selling point of reductionism. At the same time, it seems equally compelling that higher levels are often autonomous, in the sense that macro-claims can be established, corroborated, and explained without introducing micro-details. This is the benchmark of antireductionism. Can both points be maintained simultaneously? Can higher-level descriptions be both reducible to *and* autonomous from the underlying substrate? How can we use structural details to advance macro-explanations without thereby threatening their independence? The key lies in Putnam's insights. He had the right intuition when he stressed the centrality of autonomy. The mistake was turning a discussion of autonomy into a debate about reduction.

Before delving into the argument, it seems only fair to note an analogy between my proposal and an influential precursor. The possibility and

---

[3] Recall from Chapter 1 that this characterization of levels as "coarse-grained" vs. "fine-grained," "micro" vs. "macro," or "higher" vs. "lower," should be understood as relativized to a specific choice of explanandum.

importance of finding some sort of middle ground between reductionism and antireductionism has been discussed in the literature. Notably, philosophers working within the new wave of mechanistic philosophy have presented their approach as an attempt to move past the current dichotomy:

> Our project began as an attempt to understand reductionistic research in science, but we soon found that neither the classic [Nagelian] philosophical treatment of reduction [. . .] nor various modifications that followed [. . .] offered an appropriate framework for the biological research cases we had chosen to analyze. We ended up recharacterizing the science in terms of mechanistic explanations in which systems are decomposed into their component operations and dropping use of the word *reduction* altogether. (Bechtel and Richardson 2010, p. xxxvii)

As stressed in Chapter 7, I am intellectually indebted to and sympathetic with the broad mechanistic outlook. Yet, I believe that their blade hardly cuts deep enough. Mechanistic philosophy has failed to transcend old divisions, as witnessed by the observation that whether the new wave of mechanism fits in better with a general reductive or antireductionist stance is a matter of constant debate. Some see a connection, more or less explicit, with modest forms of reductionism.[4] At the same time, because of its focus on abstraction and idealization, which confer to higher-level mechanistic description an autonomy of sorts, others present mechanistic philosophy as an alternative to reduction.[5] What this shows is that mechanistic theory has not freed us from the old dichotomy. I would like to take this one step further, and move away from both stances, stressing how they can be effectively reconciled. With these clarifications in mind, let us get back to business.

Here is the punchline. The problem with reductionism vs. antireductionism debates is the more or less tacit presupposition that macro-explanations

---

[4] For instance, Glennan (2017) notes: "Because New Mechanists emphasize the ways in which the organized activities and interaction of parts explain the behavior of wholes, it might seem that New Mechanist ontology is committed to a pluralist (*and reductionist*) view" (p. 56, italics added). Glennan is adamant in stressing that his approach avoids radical "nothing-but" forms of reductionism. Still, it has a clear reductionist flavor. Similarly, Godfrey-Smith (2014) notes, "This kind of [mechanistic] work is "reductionist," in a low-key sense of that term: the properties of whole systems are explained in terms of the properties of their parts, and how these parts are put together" (p. 16).

[5] In the words of another prominent advocate of new mechanism, "I argue that the mosaic model of the unity of neuroscience, based on the search for mechanistic explanations, *is better suited than reduction* to the descriptive, explanatory, and epistemic projects for which these classic models were designed" (Craver 2007, p. 233, italics added).

and their micro-counterparts have the same explanandum, the same object of explanation. What most discussants fail to note is that, by transforming higher-level questions into lower-level ones, we are not providing different, competing answers to the same query. Inquiries framed at different levels are, for all intents and purposes, different questions. Failure to acknowledge this simple—albeit powerful—point leads to much confusion, including viewing reduction and autonomy as antithetical, when they are not. This is the fulcrum of my argument. I now need to put some meat on these bare bones.

What is the relation between hypotheses at different levels? Borrowing a Kuhnian metaphor, one could say that explanations with substantially different scope are typically "incommensurable." As we saw in Chapter 9, incommensurability was one of the main tenets of Kuhn's philosophy of science. His claim that scientific paradigms are "incommensurable" is often understood as a radical form of conceptual relativism. This, simply put, is the claim that a paradigm shift alters the language of a theory so drastically that superficially similar claims turn out not to be mutually translatable across paradigms. This, as we discussed at length in the previous chapter, is too strong. Paradigms are seldom, if ever, incommensurable in this extreme sense.

Our excursus into the nature of black-boxing draws attention to a subtler, more compelling, and less devastating form of incommensurability. This is the claim that hypotheses and explanations framed at different levels cannot be compared *directly*. They are always mediated by models, theories, and other vehicles of representation. This context-relativity boils down to a sort of testing holism, albeit one whose consequences are fairly mild. Basically, it is always possible to adjudicate between competing hypotheses directly, when these alternatives are embedded within the same model. Nevertheless, whenever one is trying to contrast explanations pertaining to different models, what is compared are not the individual hypotheses themselves, but a broader theory or paradigm. I should stress that my point is not the radical claim that hypotheses can never be compared across models. That would be tantamount to advocating a radical form of meaning incommensurability and giving up on the progress of science. My suggestion is that when one compares, say, a Newtonian explanation with its relativistic counterpart, one is not assessing hypotheses directly. Rather, in a more Duhemian fashion, one judges the explanatory forces of two entire paradigms. As we shall see in the following section, a similar relation holds when we contrast genetic vs. molecular explanations, or psychological vs. neuroscientific ones.

These considerations raise an important question. What is the relation between paradigms? How do we assess two hypotheses or explanations pitched at distinct levels or scale? Allow me to address this issue by introducing an important—if sometimes forgotten—philosophical distinction.

In his classic "Empiricism, Semantics, and Ontology," Carnap (1956a) observes, "If someone wishes to speak in his language about a new kind of entities, he has to introduce a system of new ways of speaking, subject to new rules; we shall call this procedure the construction of a linguistic *framework* for the new entities in question" (p. 206). Carnap then draws a divide between two kinds of questions concerning ontology, or what there is. On the one hand, there are questions concerning the existence or reality of entities within a framework—call these "internal" questions. On the other hand, there are questions concerning the existence or reality of the system of entities as a whole. Carnap calls these latter questions "external."

A few examples should clarify the main idea. Are numbers real? This question, Carnap maintains, is subject to two interpretations. On the one hand, it could be understood internally, that is, relative to the system of numbers of basic arithmetic. From this perspective, the answer is trivially true: of course there are numbers in the framework of arithmetic. Note, however, that not all questions of this sort are obvious. Consider, for instance, the existence of the largest prime number, or Goldbach's conjecture, the currently unresolved hypothesis that every even integer greater than 2 can be expressed as the sum of two primes. Here the answer is far from evident. Still, what all these hypotheses have in common is the presupposition of a framework, the system of arithmetic. This is what makes them internal.

On the other hand, questions concerning the existence of numbers can be understood externally. Thus conceived, in asking whether numbers are real, one is inquiring into the ontological status of numbers, that is, their "reality" independent of arithmetic or any other linguistic framework.[6] How is one supposed to adjudicate ontological questions from this external perspective? Carnap contends—correctly in my opinion—that such a question, strictly speaking, has no cognitive content. Yet, this is not to dismiss it as gibberish. The right interpretation is not as whether or not, say, numbers have the property of existence. Rather, it is a pragmatic issue of whether or not we should accept or introduce the framework of natural numbers, as opposed to a different one. The choice between frameworks is not one capable of being true

---

[6] For an excellent example of this kind of "external" inquiry, see Benacerraf (1965).

or false. In *this* sense, Carnap claims, it lacks cognitive content. It is a pragmatic matter of usefulness.

Hang on a second. Some readers may wonder whether they have missed something. Since the very beginning of this work, I have declared positivism dead and philosophy of science in need of a different guiding spirit. And, now, am I really drawing upon the work of one of the main figures in logical empiricism, namely, Rudolf Carnap? Short answer: yes, yes, and yes. When critically assessing a historical movement, we must be careful not to throw out the baby with the bath water. Logical positivism is no exception. The naïve derivational reductionism and the cumulative conception of progress ought to be abandoned. In contrast, Carnap's distinction between internal and external questions, as well as its connection to Quine's attack on the analytic-synthetic distinction, remains very important for contemporary philosophy.[7] All of this just goes to show how black boxes—and philosophical discussions thereof—are full of surprises. One can never predict in advance what is going to pop out once you open them.

So, where am I going with this? Carnap's notion of linguistic framework, and his related distinction between internal vs. external questions, can be applied to the explanatory-relativity that is integral to the black-boxing approach. Allow me to elaborate. Recall from section 5.2 of Chapter 5 that all requests for explanation presuppose a context. Once the backdrop is in place, we can ask whether the explanation is plausible, correct, if it can be amended, etc. These are all internal questions, since they occur relative to the choice of a model—or, at least, a preliminary frame. Consider a simple example from genetics. Suppose we are wondering which among two genes, call them g_1 and g_2, is responsible for transcribing protein p. This question is an empirical one, in the sense that we have two candidate answers to the inquiry: which gene transcribes p, and we need to figure out the solution.

Nevertheless, it often happens that competing explanations do not share a background theory, framework, or paradigm. Consider, for instance, alternative accounts of human behavior that appeal to genetics vs. upbringing (nature vs. nurture), or psychological vs. neuroscientific characterizations of behavior. Even if, at very general levels of description, such depictions converge on the same explananda ("inheritance," "behavior") the explanations

---

[7] Coming from a similar perspective, Bickle (2003, Ch. 1) resurrects a Carnap-inspired internal-external distinction as part of meta-science. For a discussion of the revived importance of Carnap's distinction in contemporary ontology, see Chalmers et al. (2009).

themselves cannot be compared *directly*, as they are framed quite differently. As such, they may not share much in terms of precisely what patterns of events they are trying to capture, as well as their standards and presuppositions. In such cases, we need to compare entire models, not individual explanations. In this sense, these explanations are incommensurable: they cannot be compared directly; they require the mediating role of a model. But models are useful or useless, clear or obscure, illuminating or not. They are not true or false. Hence, the choice between models is not a cognitive one, in Carnap's sense. It is an external question, a choice involving which framework to adopt. Note how well this fits in with the "diet" mechanism of Chapter 7, according to which whether a system is depicted as an entity or as a box inherently depends on the choice of model.

In conclusion, reductionists and antireductionists share the presupposition that explanations pitched at different levels provide competing explanantia of the same explananda. This is a mistake. They do not. Explanations are context-relative. It should now become clear in what sense higher-level explanations can be both reducible *and* autonomous from lower levels. Reduction is a relational issue, concerning the reframing of a macro-explanandum as micro-inquiry. Autonomy is an intrinsic property of an explanation, pertaining to whether a model provides the resources to adequately address its explananda. Thus conceived, these two properties are perfectly compatible.

Let me illustrate the coexistence of autonomy and reduction by focusing on our toy example. Scientific scenarios will be covered in section 10.4.

In Putnam's square-peg round-hole thought experiment, the core issue is whether or not structural physical details are relevant to the explanation of the behavior of the system and, if so, whether they increase explanatory power. Unless we clarify what it is that we are trying to explain, such a question is meaningless. If the explanandum is the observation *that* the square peg does not pass through the round hole, then the physical details can—indeed, must—be effectively black-boxed. In contrast, if the objective is explaining *why* this state occurs, then looking at the physical structure of the system, down to its atomic constituents, becomes relevant. In this case, the black-box must be opened and replaced by a mechanism.

Putnam's deep insight was recognizing that higher-level explanations can be epistemically autonomous from lower-level ones. His error, which turned out to be quite consequential in the wake of the ensuing debate, was to spin this as a vindication of antireductionism. Putnam, like many other philosophers thereafter, identified autonomy with antireductionism. He was wrong. These concepts should not be conflated. The reason is that, properly conceived, the reductionism vs. antireductionism debate involves external questions *concerning* the choice of a model and levels of explanation. Autonomy, in contrast, underlies an "internal" question concerning the independence of a causal explanation *given* the choice of a model.

§10.4. A History of Black Boxes

Section 10.3 argued that the black-boxing strategy reconciles autonomy and reduction, the signature traits of antireductionism and reductionism, respectively. Let us see how this works in practice by revisiting, one final time, the case studies that accompanied us throughout our journey together.

§10.4.1. Beyond Reductionism and
Antireductionism: Evolution

To get started, take one last look at Darwin's explanations. Recall that his explicit target is the distribution of organisms and traits around the world. His evolutionary explanans, simply put, is descent with modification, fueled by natural selection. Is this a story of reductionism or antireductionism?

Here is a reductionist rendition of the tale. Darwin is surely correct that descent with modification is the central factor in explaining distributions of organisms and traits across the globe. Evolution by natural selection is, indeed, the principal frame and difference maker. But how should we understand evolution by natural selection? Darwin himself breaks down the process into four key ingredients: variation, competition, fitness, and heritability. This was an important insight. But it was only the beginning. In the wake of Darwin's groundbreaking work, progress was gradually achieved by decomposing these broad concepts into more fundamental components. This is precisely what one would expect from a reductionist perspective.

To further elaborate, consider Sober's example of frequency changes in a population of *Drosophila*. Initially, variation at the population level may be explained by positing that type *A* flies are "fitter" than type *B* ones. As the system is studied further, details emerge. In the imaginary case at hand, it turns out that the fitter type *A* is characterized by a chromosome inversion that produces a thicker thorax, better insulating the organism, which makes it more resistant to cold weather. As Sober notes, at this point, appeals to "fitness" become disposable. Fitness attributions can be replaced by descriptions of the mechanisms producing the relevant frequencies. And, indeed, this more precise depiction provides a deeper account of changes at the population level. In short, "fitness" is explanatory. But describing the mechanisms responsible for differential survival is *more* explanatory. And further decomposing these genetic mechanisms into more fundamental constituents will make the model even *more* powerful. This downward trajectory, the reductionist says, is captured by the reductive outlook.

Not so fast, the antireductionist retorts. Sure, the chromosomal description is more detailed than the preliminary ascription of fitness. Still, is this, in and of itself, a vindication of reductionism? Accuracy is an important aspect of explanation. But it is not the only factor at play. Shifts between levels involve a trade-off in explanatory power. Micro-descriptions are more precise; macro-depictions are more general. To wit, what do the fittest *Bacteriophage* λ, the fittest *Drosophila melanogaster*, and the fittest *Homo sapiens* have in common, which sets them apart from other members of their species? Good luck answering this question by pinpointing shared biochemistry. And insisting that, from a biophysical perspective these organisms have *nothing* in common just goes to show that moving down on our levels of descriptions is a compromise. Some features are gained, while others are lost. In short, lower-level depictions are not invariably more powerful than higher-level ones. They merely focus on different aspects. Fitness is just an instantiation of a broader phenomenon. Darwin's explanations in the *Origin*, which are still widely accepted by evolutionists, vindicate the antireductionist story.

At this point, reductionists will presumably point out all the contributions of molecular biology, broadly construed, to the study of evolution. Sophisticated antireductionists should agree. But they also maintain that what reductionists insist in dubbing "reduction" is really a case of multilevel integration. Modest reductionists, in turn, will hold their ground, claiming that this is all part of a molecular basis. And—here we are, back right where we started! Both parties recognize Darwin's remarkable contributions and the progress that has been made since. The disagreement is whether this should be recounted as a story of "reductionism" or "autonomy."

How about trying something different? Consider the situation from the perspective of black-boxing. First, note that variation, competition, fitness, and heritability are placeholders in causal explanations framed in models—they are black boxes. The construction of a black box begins by sharpening the object of explanation. Next, one determines what brings about this explanandum and embeds the causal story within a model. The significance of the framing stage is that macro-explanations and their micro-counterparts have distinct explananda, which presuppose *sui generis* explanantia. If the target is an explanation of frequency changes in the fruit fly population, fitness provides a perfectly adequate explanans. What fitness alone does not explain is why type *A* flies are more resistant than type *B* flies. This is an important question, albeit one that frames a very different explanandum. From this perspective, reductionists are correct in stressing that molecular biology deepens our evolutionary accounts. At the same time, this is not to reject the autonomy of evolutionary biology. This is because explanations are relativized to contexts. As a result, we cannot compare them directly. What is compared are entire models, not individual hypotheses.

In sum, does the story of Darwin vindicate reductionism or antireductionism? Both or, perhaps, neither. Either way, it hardly matters. Reconstructing the situation in terms of black boxes shows that micro- and macro-explanations are *both* autonomous *and* reducible. They are autonomous because, when appropriately framed, higher-level depictions stand alone, independent of structural details. At the same time, higher-level accounts are reducible because there are models that provide more detailed and more powerful lower-level depictions. The mistake, common to both reductionists and antireductionists, was assuming that these two stances are incompatible. The root of the trouble is presupposing that the two explanantia have identical explananda. They do not. Asking whether lower-level explanations are "better" than higher-level ones is an ill-posed question. The explanations are incommensurable. What is compared are not individual hypotheses, but the models, the paradigms that embed them. This comparison is not a matter of truth or falsity. It corresponds to what Carnap dubbed an "external" question, to be addressed on purely pragmatic grounds.

§10.4.2. Beyond Reductionism and Antireductionism: Genetics

Our second case study focuses on the development of genetics, from Mendel's pioneering insights to the contemporary landscape. Originally showcasing the limitations of Nagel's account, the relation between classical and molecular genetics has grown into a poster child for antireductionism.

Mendel's explicit targets are the observable inheritance patterns, sometimes referred to as "Mendelian ratios." Each organism, Mendel hypothesized, inherits two "factors"—subsequently called "genes"—one from each parent. When these factors are different, one is always expressed preferentially over the other. These factors are passed on, unchanged, to the next generation. Why would this well-known story support antireductionism?

To restrict the scope of the discussion, consider Mendel's second law, the "law of independent assortment," which, simply put, states that genes located on non-homologous chromosomes assort independently. This generalization, as Kitcher maintains, is best explained at the cytological level. To be sure, the process of meiosis can be described, in much greater detail, in biochemical terms. But does this knowledge enhance the original cytological depiction? Antireductionists answer in the negative.

Many reductionists disagree, complaining that this perspective is simply outdated. Molecular biology has greatly improved the study of gene replication, expression, mutation, and recombination. Insisting that explanatory power is, in principle, immune from lower-level revision and enrichment is simply wrongheaded. Mendel's insight was on the right track. But much remained to be done. What is the physical structure of a gene? How are these traits inherited, transmitted, and expressed? We now have a much deeper understanding of genes and other genetic processes. Mendel's insights have not just been vindicated. They have been expanded and explained.

The problem with this response, from an antireductionist perspective, is that it sets up a straw man. Of course, biochemistry and other molecular insights enhance our knowledge of genetics. This was already acknowledged in Kitcher's pioneering "1953" article, where molecular biology is presented as an "explanatory extension" of classical genetics. The point, stressed by antireductionists ever since, is that higher-level natural kinds do not count as natural kinds at all from a lower-level standpoint. Recall how the classical Mendelian concept of a gene was supposed to fulfill three roles. It was intended as the unit of mutation, the unit of recombination, and the unit of function. No single biochemical entity fulfills all three. Hence, reductionism distorts the success of classical pre-molecular genetics. We need an antireductionist outlook to do justice to these groundbreaking discoveries.

Now it is time for reductionists to complain. Multiple-realizability did pose an inescapable trap for classical reduction, which required a series of type-identities across levels. New-wave epistemic reductionism, in contrast, does not fall prey to this objection. What if no single biochemical unit corresponds to Mendelian genes? Is it just that the microcosm is more complex than Mendel himself and his early twentieth-century followers could surmise? Furthermore, of course we need to employ functional language to correctly study genes, chromosomes, and the like. Molecular biology is not an attempt to purge biology of all talk of teleology. *Au contraire*, it is an attempt to "naturalize" the field, by uncovering the entities, processes, and activities that instantiate these functions and dispositions. This process enhances explanatory power, vindicating the reductionist maxim.

It should be clear by now where the discussion has gone astray. The issue is not whether biochemistry can shed light on genetics, evolution, or the rest of biology. It does, and everyone is well aware of this. The question underlying the debate is whether the relation between classical and molecular genetics better conforms to the reductionist or the antireductionist paradigm. And, as noted in section 2.6 of Chapter 2, this ultimately hinges on the definition of "purely molecular language" and other terminological conventions.

Let us try reassessing the situation from a different perspective. Genes and other genetic constructs are black boxes. They are placeholders in causal explanations framed and represented in models. In this regard, reductionists are correct in pointing out that biochemistry provides much deeper and more insightful analyses, compared to the relatively shallow accounts of classical genetics. Thus conceived, the realization that, from a biochemical standpoint, the Mendelian gene does not exist as a single structure is a vindication of the reductionist strategy, not a failure thereof. At the same time, this does not entail a rejection of autonomy at higher levels. We still teach Mendel's laws to beginning biology students. This is because these generalizations capture important biological facts and convey them well, striking the right balance between precision and generality.

In sum, reductionism and antireductionism are both correct, but not exclusively so. Reductionism rightly notes that the molecular paradigm provides deeper, more accurate, and more detailed explanations. The insight of antireductionism is the autonomy of higher-order paradigms. Both parties ultimately miss the mark by tacitly presupposing that these explanations are directly in competition with one another. They are not. Strictly speaking, the explanations are incommensurable. This, however, does not mean that molecular biology is incomprehensible from the perspective of cytology, or vice versa. The point is that what is compared is not individual hypotheses, but entire theories, frameworks, paradigms. And the choice between them is "external," that is, pragmatic. Thus construed, Mendelian genetics is *both* autonomous *and* reducible to molecular biology.

§10.4.3. Beyond Reductionism and Antireductionism: Behaviorism

Radical behaviorism in psychology provides a different, interesting case study for a simple reason: it is a story of failure. Virtually no contemporary scholar, in psychology or philosophy, flatly identifies with a full-fledged behaviorist, at least when behaviorism is conceived as a theory of mind. Consequently, the question underlying Skinner's ultimately unsuccessful attempt to provide new foundations for the field of psychology becomes: was he guilty of a reductionist mistake or an antireductionist one?

Skinner's target is a complex reconstruction of psychology that eschews any appeal to the internal mental states of agents. As noted in section 6.4 of Chapter 6, Skinner's rejection of mentalism is more complex than is often assumed. The crux of his argument can be phrased in terms of the theoretician's dilemma. If, on the one hand, mental states link environmental stimuli and behavior in a deterministic fashion, the story goes, mental states become completely unnecessary for the explanation of behavior. If, on the other hand, the mental-state-mediated links between stimuli and behavior are indeterministic, the mental states are utterly useless. Either way, references to mental states do not contribute to psychological explanations of behavior and should therefore be omitted. Environmental inputs are all we really need.

Skinner's theoretician's dilemma does not hold up to serious scrutiny. It is controversial, to say the least, that deterministic links of the form $E \to M \to B$ make appeals to M unnecessary. After all, even in a fully deterministic universe, M could provide better, more accurate explanations of behavior than E. This first horn of the dilemma requires further argument to become convincing at all. But the real problem emerges when we shift to the second horn. Why would indeterministic links make M disposable? Non-deterministic generalizations may be perfectly explanatory, as witnessed by their extensive use throughout the sciences, from quantum mechanics to evolutionary theory, from biochemistry to economics.

In short, Skinner's negative argument against mentalism leaves us wanting. Moreover, all positive evidence points to an indispensable role of mental states, broadly construed, within psychology. Skinner was wrong on this score, or so it seems. But what moral should one draw from his mistakes?

Antireductionists will likely blame reduction. Skinner's shortcoming lies in his attempt to reduce mental states to complex behavioral dispositions. Interestingly, from this perspective, the error is pervasive across psychology and the philosophy of mind. Behaviorism was hardly the only reductive theory of mind developed in the century just passed. U. T. Place and J. J. C. Smart's identity theory purported to reduce mental states to brain states, a view subsequently developed by Francis Crick's "astonishing hypothesis" (1994) Eliminative materialism, a position advanced and advocated by Patricia and Paul Churchland, maintains that mental states are theoretical entities posited by folk psychology. Widespread and commonsensical as it is, folk psychology, they argue, is a flawed theory of mind and, as such, it is not a candidate for integration. Rather, it should be eliminated and replaced by a mature neuroscience, which will be more predictive, more explanatory, and better connected to extant research. Despite evident differences, from an antireductionist perspective, behaviorism, identity theory, and eliminative materialism partake of the same root problem: they purport to reduce or replace mental states with something else. Contemporary neuroscientists who follow suit and identify mental states with brain states are committing the same kind of mistake as Skinner, and their proposals are doomed for precisely the same reasons. The solution is to adopt an altogether different theory of mind, one that does not dispense with mentality. Functionalism, token-identity theory, and property dualism are all popular candidates.

Reductionists will surely provide a very different reading of the situation. The trouble with radical behaviorism, this story goes, has nothing to do with reduction per se. The mistake lies in the type of reduction. There is more to mental states than behavioral dispositions. The key to mind-body reduction is an appropriate neuroscientific foundation for psychology.

In sum, the quest for the nature of mind has turned into a battleground for reduction. Can mental states be reduced to physical states? If so, how? If not, why not? This reducibility of mental to physical states, or impossibility thereof, lies at the core of twentieth-century philosophy of mind.

If reductionism is so central to contemporary philosophy of mind, we asked back in section 2.6 of Chapter 2, where do we stand with respect to psycho-neural reduction, after decades of extensive debate? As noted, not too much success can be boasted. To be sure, much progress has been achieved in the unraveling of psychological and neural mechanisms underlying higher and, especially, lower cognition. But has this advanced the philosophical debate over the reduction of mind? If so, the news has not yet been broken, as there seems to be no more consensus on this matter than there was in the 1950s.

It is worth recalling that, when confronted with this lack of resolution, scientists and philosophers alike tend to respond by pointing to the complexity of the human brain, with its astronomical number of connections among billions of cells. I obviously do not question how hard it is to study brains and minds. I am, however, skeptical that the complexity of the underlying structure is responsible for lack of tangible progress. The main culprit is the notion of (anti)reductionism itself. This should be old news. What's novel is that we can now finally see how to overcome the false dichotomy.

Consider behaviorism from the perspective of black boxes. Watson and Skinner's attempt to place psychology on more secure methodological grounds had a momentous impact on the entire field. First, it stressed the tight connection between mental states and behavior. Second, and more important, it contributed to making psychology more "scientific." Their insight to black-box mental states was indeed revolutionary. By setting psychoneural patterns and mechanisms aside, Skinner and colleagues were able to draw attention to the importance of stimuli, operant conditioning, and other forms of environmental effects on human conduct. At the same time, radical behaviorists believed that the content of this black box had no substantial place in psychology. This was their crucial mistake. In the wake of the cognitive revolution, it became increasingly clear that mental dispositions, and other psychological states, play an irreplaceable role in the explanation of human conduct. As we saw, Watson did embrace a simplistic and naïve reductionism. Not so Skinner, who was guilty of neither reductionism nor antireductionism, which are external, pragmatic stances. His error is best understood in terms of the black-boxing strategy. His explanandum is framed incorrectly. Consequently, the difference makers are misidentified, and the ensuing causal model of behavior is lacking. All of this is independent of reductionism and autonomy. Skinner's shortcomings lie in the details of his model of behavior—the construction of his black box.

§10.4.4. Beyond Reductionism and
Antireductionism: Economics

The heart and soul of neoclassical economics is the attempt to predict and explain economic behavior on the basis of choice-related data. The ingredients for this simple, yet controversial, recipe include formal postulates like the weak and general axioms of revealed preference ("WARP" and "GARP"), as well as the ideal of rationality at the core of Expected Utility Theory.

The past few decades have witnessed the rise of alternative approaches, which I subsumed under the moniker "psycho-neural economics." The basic insight is that details concerning how human minds and brains actually frame, compute, and resolve problems challenge fundamental assumptions regarding the behavior of agents. For instance, behavioral economists stress how people rely heavily on heuristics, biases, and reference points when adjudicating potential outcomes, as opposed to computing expected utilities. And, clearly, from the standpoint of psycho-neural economics, heuristics, biases, and reference points are not the means by which rational agents approximate the predictions of expected utility theory. Certainly Friedman, or someone of his behalf, could make the argument that psychological data of various kinds do not disconfirm or otherwise call into question his "as if" approach.

An informed prediction of the trajectory of economics lies beyond my interest and professional competence. The important point, for present purposes, is that the debate between neoclassical and psycho-neural economists can be reconstructed along the reductionism vs. antireductionism lines. But *should* it? Unsurprisingly, I offer a negative answer and an alternative.

The reductionist story goes like this. Characterizing utility in terms of revealed choice was an effective strategy early in the twentieth century, which allowed economics to rest on more secure methodological grounds, making it kosher from a scientific—read: positivist—standpoint. Specifically, it allowed economists to transform utility ascriptions from spooky unquantifiable mental states to empirically testable hypotheses. This effective strategy faced some shortcomings. In particular, it was founded on a series of "as if" models, introducing various unrealistic psychological assumptions. This was a necessary move decades ago, when little was known about how this information is actually processed. The landscape, however, has changed drastically. We now know a lot more about the psychological mechanisms underlying human decision-making. Consequently, economics can advance by replacing "as if" models with more realistic psycho-neural assumptions.[8]

Given the methodological ties linking early neoclassical economics to positivism, which explicitly endorses the in-principle reducibility of all of science to physics, it might seem paradoxical to present this branch of economics as a form of antireductionism. Nonetheless, from our contemporary perspective, Friedman's influential methodology of positive economics bears a striking resemblance to antireductionist arguments supporting autonomy.

Friedman defends the foundations of neoclassical economics from the charge that it presupposes hypotheses known to be false. His discussion of the law of free-falling bodies, leaves on trees, and professional billiard players are intended to put economics on a par with other natural sciences, as well as common sense. We know all these hypotheses to be false. And, yet, everyone accepts them as unproblematic. Then why the fuss about individual firms behaving as if they strive to maximize profit? Truth or falsity of assumptions, Friedman famously concludes, is irrelevant with regard to the acceptance or rejection of a scientific theory. What really matters is whether the hypothesis is predictive and explanatory with respect to the class of phenomena that it is intended to cover.[9] Hence, the relevant question to ask about the assumptions of a theory is not whether they are descriptively realistic. They never are. The issue is whether the hypotheses in question are good enough approximations for the purposes at hand.

These analogies spell out a powerful argument in favor of the autonomy of economics from neighboring disciplines. Economics is, in principle, shielded from "lower" sciences, psychology and neuroscience included. In the context of neoclassical economics, choice is assumed, not explained. But is it really true that psychological discoveries cannot have any impact on economics?

From the standpoint of black-boxing, both stories capture something important regarding the relation between economics, psychology, and neuroscience. And both leave out a significant piece. Neoclassical economists are right to stress autonomy. Economics is neither psychology nor

---

[8] Some neuroeconomists, such as Glimcher (2011), are explicit and unapologetic about their reductive endeavors. Other authors prefer to replace talk of "reduction" with the "integration" of psychology, economics, and neuroscience (Craver and Alexandrova 2008). Setting aside differences among these strategies, the central point is that it is important to reconcile the claims of economics with the best theories and data from psychology and neuroscience, as these fields have the potential to mutually inform each other.

[9] Recall from section 4.4 of Chapter 4 that, following the deductive-nomological model, Friedman treated prediction and explanation as two sides of the same coin.

neuro-science. These disciplines have different objectives and demand different abstractions. But completely insulating economics from these fields is unreasonable. Psycho-neural economists correctly note that psychology and neuroscience may inform economics in various ways. The challenge is how to reconcile these seemingly incompatible claims. How can we make psychology and neuroscience relevant to economics while maintaining the methodological autonomy of these disciplines? Black boxes to the rescue.

The solution is to construct utility as a black box. At a general level of description, utility can be used to frame the notion of economic behavior, which brings the two theories together. Neoclassical and psycho-neural economics both study Robbins's "human behavior as a relationship between ends and scarce means which have alternative uses" (Robbins 1932, p. 16). But when we start unpacking this black box, we find very different frames and difference-makers. Psychology and neuroscience do a fine job unraveling the mechanisms underlying decision-making. Yet, in doing so, they address very different explananda, incommensurable to economic ones, which makes the choice between them "external." The integration of psychology and economics cannot be achieved by uncovering underlying mechanisms. It rather depends on finding broader, overarching frameworks that construct black boxes spanning both fields.

§10.4.5. Beyond Reductionism and Antireductionism: Phlogiston

Before moving on, let me briefly address our final example: the case of phlogiston. This is another peculiar case study since, differently from the others, it involves an entity that has been purged from our scientific ontology. This raises a host of distinctive questions and challenges.

Recall how phlogiston theory purported to provide an explanation of the process of combustion. What happens when a log is burned? What is the nature of the reaction that turns wood into ashes? What accounts for changes in mass, alterations to the surrounding air, and so forth? The keystone of phlogiston theory is the postulation of an entity—phlogiston—that is supposedly emitted in all cases of combustion. This concept, we now all know, is not a natural kind. There is no substance that is emitted in all cases of combustion. There is no such thing as phlogiston. Still, as we saw in Chapter 9, the progress of the theory itself raises interesting conundrums. The question that I want to address here is this: is the transition from phlogiston theory to atomic chemistry a reductionist or an antireductionist story?

From a radical reductionist perspective, the natural conclusion to draw is that phlogiston has been eliminated or reduced to concepts and entities postulated by modern atomic chemistry. The causal role played by phlogiston in the old theory is now performed, for the most part, by the oxidation of organic compounds. The problem, in brief, is that this seems to throw out the baby with the bath water. Sure, we can all agree that there is no phlogiston out there in the world. Still, phlogiston theory did make strikingly accurate predictions. How does a nonexistent substance explain anything at all? How do we account for the partial success of the theory?

Traditional antireductionism provides a diametrically opposite perspective, according to which phlogiston theory has been rephrased in terms of modern chemistry. This solves the issue of success, which undermines the eliminativist story. If what Priestley and colleagues called "phlogiston" actually refers to, say, oxygen and other chemical elements, then there is no mystery as to why appeals to phlogiston are partially explanatory. But we now face trouble explaining why the theory came to be discarded at all. Furthermore, if not even phlogiston counts as an instance of elimination, does any theoretical entity in science ever come to be thrown out?

In short, uncompromising forms of both reductionism and antireductionism miss the mark. Case histories like this one require a more nuanced and subtle middle ground that recognizes how parts of a theory have been eliminated or reduced, whereas other parts have been integrated and rephrased in contemporary terms. The question is: which parts and why?

On Kitcher's view, theory-laden terms, such as "phlogiston," have a complex reference potential. The meaning and denotation of these expressions is context-dependent: they may refer to a plurality of entities, processes, and activities, as well as to nothing at all, depending on the circumstances and the intentions of both the speaker and the surrounding linguistic community.

Chapter 9 argued that Kitcher's notion of reference potential and my blackboxing strategy mutually reinforce each other. When does "phlogiston" refer to oxygen? When does it pick out something else? When does it flatly fail to refer? The key to answering these questions is to note that theory-laden expressions like "phlogiston" are embedded in a rich and complex theoretical thicket. As such, their framing is crucial. This process singles out the presence or absence of phlogiston as the crucial factor in the process of combustion.

This is the difference-making stage, which then requires the representation of these causal factors within a model.

In brief, the intentions of speakers and the presuppositions of linguistic communities surely play a crucial role in the determination of reference. But understanding whether the term in question is employed as frame or difference-maker, and how it is represented in the context of a model, is also of paramount importance for determining when the term "phlogiston" picks out a behavior in need of explanation, whatever is responsible for producing the behavior in question, or when it refers to nothing at all.

In sum, the story of phlogiston combines aspects of reductionism with aspects of antireductionism. Reductionists are correct that progress was achieved by breaking down phlogiston and rephrasing its explanatory successes in terms of atomic chemistry. At the same time, antireductionists rightly stress the autonomy of phlogiston theory, whose explanatory success—or lack thereof—should be assessed on its own grounds. Once again, all of this goes to show that autonomy and reduction are perfectly consistent. The reason is that what we compare and assess are theories, not individual hypotheses.

§10.5. The Devil in the Details?

Time to take stock. There is a common misconception underlying both traditional reductionism and classic antireductionism. This is the assumption that the same questions can be raised at different levels. On the contrary, inquiries are framed differently and thus do not correspond across theories, fields, and paradigms. In this sense, they are *incommensurable*. Physical descriptions do not answer the same questions posed in chemistry, biology, or higher steps in the hierarchy. These inquiries are context-dependent. As such, we cannot compare them directly. What we compare are entire models, and the choice between frameworks themselves is a Carnapian "external" question. The bottom line is that theoretical progress is not a matter of reductionism vs. antireductionism or decomposition vs. autonomy. It is both, and much more. The key to the advancement of science is a matter of framing the right inquiries, identifying the appropriate difference-makers, and constructing useful models. In a nutshell, it is a matter of packing and unpacking black boxes.

What becomes of the old methodological dichotomy between Scylla and Charybdis? As discussed at length throughout the book, the position of modest reductionists and sophisticated antireductionists overlap. Still, this does not mean that there are no controversial issues or substantial debates at play. But these disagreements are better captured in terms of black boxes, as opposed to drawing a less-than-substantive contrast between reduction and autonomy. Here is a suggestion. Many philosophers with reductionist inclinations clearly acknowledge that black boxes play a significant role in science. Yet, they contend that their function is merely a temporary convenience. Black boxes, from this perspective, are disposable scaffoldings. As science marches on, these placeholders are gradually eliminated. The regulative ideal of science is an unqualified theory of everything, expressed in an ontologically fundamental physical language—whatever that may look like—where each black box has been opened and replaced. Antireductionist supporters can retort that boxes have a more permanent status: placeholders are here to stay. From this standpoint, our best scientific models, including our most fundamental physical descriptions of the universe, necessarily include black boxes, and other forms of abstraction and idealization. A comprehensive theory of everything purged of boxes and other higher-level placeholders is neither achievable nor desirable. This contrast captures some core aspects of the old debate between reductionists and antireductionists. At the same time, in my humble opinion, such paraphrase in terms of the status of black boxes in scientific practice is clearer and more perspicuous than the old muddled question: can physics explain everything?

Before bidding farewell, I still have one last promise to fulfill. All the way back in Chapter 1, I criticized the cumulative model of science, and the corresponding way in which science is taught, on the grounds that it leaves no room for ignorance. The antireductionist pathway does not provide a promising alternative. This book spelled out a middle-ground solution that, I promised, would fare better on this score. Can I back this up? Does the black-boxing strategy shed light on the productive role of ignorance without giving up on the insights of reductionism?

All ignorance is equal, but some ignorance is more equal than other. Setting Orwellian references aside, back in Chapter 1, I briefly introduced Firestein's distinction between two kinds of ignorance. On the one hand, there is "willful stupidity," a stubborn devotion to uninformed opinion and other situations where "the ignorant are unaware, unenlightened, uninformed, and surprisingly often occupy elected offices." At the same time, Firestein observes, there is a different, less-pejorative sense of ignorance. This is knowledgeable ignorance, perceptive ignorance, insightful ignorance.

I would like to further elaborate on this distinction, by suggesting that there are not two, but *three* kinds of ignorance at play in science, as well as in ordinary settings. First, there are situations which correspond to Firestein's "willful stupidity," where someone stubbornly resists getting acquainted with the relevant, known facts. Call this *disruptive ignorance*, because this kind of ignorance has no positive role whatsoever to play, in science as in everyday life. Now, let us focus on knowledgeable, perceptive, insightful ignorance. It is important to appreciate that this idea can be cashed out in two distinct ways. First, there are situations where we cannot provide an adequate account of an event simply because we do not know enough about the event in question. But realizing what we do not know may provide some steppingstones toward acquiring the knowledge in question. Call this *weakly productive ignorance*. Contrary to disruptive ignorance, weakly productive ignorance is blameless. It is not that we refuse to acquire the relevant information; the relevant information is just not there. As Firestein notes, when properly harnessed, this kind of ignorance may become a portal to knowledge, to success. It paves the way toward more knowledge. In this sense, it is productive. There is, however, a third, less obvious kind of productive ignorance, which I call *strongly productive ignorance*. This is an instance where ignorance is not only blameless, harmless, and potentially conducive to knowledge, as in the case of weakly productive ignorance. With strongly productive ignorance, the state of ignorance may actually be preferable to the corresponding state of knowledge. Ignorance actually becomes a goal in itself. In sum, we can all agree that disruptive ignorance is bad and productive ignorance is good. Yet, productive ignorance comes in two varieties, weak and strong.[10]

The reason why this matters is that weakly productive ignorance fits in well with all extant models of science. Antireductionists and reductionists alike should agree that what we do not know may become an incentive to do better, to learn more, to accomplish what we are striving for. When I claimed, back in Chapter 1, that the reductionist image of science as erecting a wall

---

[10] These considerations, *ça va sans dire*, hardly exhaust the deep and extensive issue of ignorance. Another interesting question concerns whether there is ignorance that is, in principle, insoluble. In other words, are there questions that we do not currently know, and there is no way of knowing—something akin to Emil du Bois Reymond's *ignoramus et ignorabimus* mentioned in Chapter 1? Or do all human questions have an answer, at least in principle? This is a fascinating debate, albeit one tangential to the scope of this work.

or tiling a mosaic leaves no room for ignorance, I had a different notion of productivity in mind. This is what I now call "strongly productive ignorance." From a cumulative, brick-by-brick perspective, the very idea that ignorance may be preferable to the corresponding state of knowledge makes no sense. The bigger the wall, the better! Every time we add something we previously did not know, we have made progress. This, however, comes at a cost. Without strongly productive ignorance, it becomes hard to see how any higher-level explanation can be truly autonomous.

It should come as little surprise that antireductionist approaches fare much better on this score. By stressing autonomy, traditional antireductionism has no problem recognizing the productive role of ignorance, which goes beyond the truism that some explanations do not require further details to stand alone. What makes this ignorance *strongly* productive, is that, by adding further detail we may end up losing generality and muddying the waters. Yet, as discussed at length, antireductionism suffers from its own shortcomings. It has trouble accounting for why adding information does not invariably enrich an explanation. After all, reductionists retort, do we not make progress when we replace ignorance—negative or productive—with knowledge? Do we not learn something that may matter now or in the future? Is the very idea of ignorance being preferable to the corresponding state of knowledge not an oxymoron?

The problem with extant debates, as we saw, is that reductionists and antireductionists alike tend to share a restrictive conception of the interface between levels of explanation. The relation between microcosm and macrocosm is much richer, and more interesting, than it is standardly depicted.

The black-boxing strategy allows us to maintain the positive features of both approaches, while avoiding the respective shortcomings. Ignorance has a constructive, strongly productive role to play. It promotes autonomy, as antireductionists point out. At the same time, opening a black box always constitutes progress, as emphasized by reductionists. The key is noting that this operation transforms the old inquiry into a new problem. Opening a black box introduces new explananda without necessarily threatening the integrity or autonomy of old ones. Medawar's notion of overtaking the forces of ignorance can be maintained without sacrificing the true autonomy of higher levels. Thus characterized, the very idea of a strongly productive ignorance is not an oxymoron, after all. It plays an important role in research.

These considerations suggest some respects in which we can do better at presenting science to the younger generations. The dusty image of science as a slow, painstaking accumulation of truth has dominated the scene for the better part of the past century, and it is still dominant in textbooks. The anti-reductionist image of a dappled universe might not be the appropriate substitute. Black-boxing suggests a different image. How about recharacterizing the scientific world as a set of *matryoshka*, that is, nested Russian dolls (Figure 10.1)? This suggests that the relation between explanatory levels is neither one of layer-cake-style progressive reduction nor one of complete autonomy. The right metaphor is one of *dynamic containment*, where each discipline constitutes a system in and of itself that, however, is constrained by what goes on "outside" and, in turn, constrains what goes on "inside." This interplay, I believe, is an important message to convey to young students of science, in books as in class, and the educated public in general. What goes on at lower levels does, indeed, constrain what happens at higher floors. But the converse is also true. Framing the right macro-explananda is instrumental for raising the appropriate micro-questions. The general lesson to be learned—and taught—is that the advancement of science requires cooperation, not elimination.

In conclusion, this book sets the foundations and explores the boundaries of what is—hopefully—a fecund and exciting research project. What I have done here barely scratches the surface of a deep, unfathomed ocean. Much remains to be done to adequately study all the workings and implications of

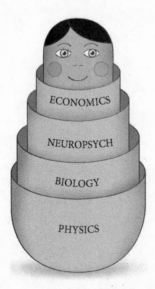

Figure 10.1. The *matryoshka* model of science.

black boxes and related concepts. If my presentation inspires others, especially younger scholars and students, by showing how important historical, scientific, and philosophical questions acquire new significance when recast and addressed in the context of this strategy, my goal will have been met in full. After all, the devil might not always be in the details. But as Goethe reminds us in the quote selected as the epigraph for this final chapter, it may be hard to catch him anyway.

References

Achinstein, P. (1968). *Concepts of Science*. Baltimore, MD: Johns Hopkins University Press.

Allen, G. E. (1975). *Life Science in the Twentieth Century*. Cambridge: Cambridge University Press.

Allen, G. E. (1978). *Thomas Hunt Morgan: The Man and His Science*. Princeton, NJ: Princeton University Press.

Amundson, R. (1983). "E.C. Tolman and the Intervening Variable: A Study in the Epistemological History of Psychology." *Philosophy of Science* 50, 268–282.

Amundson, R. (1986). "The Unknown Epistemology of E.C. Tolman." *British Journal of Psychology* 77, 525–531.

Anderson, J. (1938). "The Problem of Causality." *Australasian Journal of Psychology and Philosophy* 16, pp. 127–142.

Appiah, K. A. (2017). *As If: Idealizations and Ideals*. Cambridge, MA: Harvard University Press.

Barnes, E. (2012). "Emergence and Fundamentality." *Mind* 121(484), 873–901.

Batterman, R. (2002). *The Devil in the Details: Asymptotic Reasoning in Explanation, Reduction and Emergence*. Oxford: Oxford University Press.

Bechtel, W. (2011). "Mechanism and Biological Explanation." *Philosophy of Science* 78, 533–557.

Bechtel, W., and R. C. Richardson. (2010). *Discovering Complexity: Decomposition and Localization as Strategies in Scientific Research* (2nd ed.). Cambridge, MA: MIT Press.

Bedau, M. A. (1997). "Weak Emergence." In J. Tomberlin (Ed.), *Philosophical Perspectives: Mind, Causation, and World*, Vol. 11, pp. 375–399. Malden, MA: Blackwell.

Bedau, M. A. (2002). "Downward Causation and Autonomy of Weak Emergence." *Principia* 6, 5–50.

Bedau, M. A. (2012). "Weak Emergence and Computer Simulation." In P. Humphreys and C. Imbert (Eds.), *Models, Simulations, and Representations*, pp. 91–114. New York: Routledge.

Behe, M. J. (1996). *Darwin's Black Box: The Biochemical Challenge to Evolution*. New York: Free Press.

Benacerraf, P. (1965). "What Numbers Could Not Be." *Philosophical Review* 74, 47–73.

Bickle, J. (1998). *Psychoneural Reduction: The New Wave*. Cambridge, MA: MIT Press.

Bickle, J. (2003). *Philosophy and Neuroscience: A Ruthlessly Reductive Account*. Dordrecht: Kluwer.

Bogen, J. (2008). "Causally Productive Activities." *Studies in History and Philosophy of Science* 39(1), 112–123.

Boniolo, G. (2013). "On Molecular Mechanisms and Contexts of Physical Explanation." *Biological Theory* 7(3), 256–265.

Boring, E. G. (1950). *A History of Developmental Psychology* (2nd ed.). New York: Appleton-Century-Crofts.

Box, G. E. (1976). "Science and Statistics." *Journal of the American Statistical Association* 71(356), 791–799.

Burge, T. (2013). "Modest Dualism." In *Cognition through Understanding: Philosophical Essays*, Vol. 3, pp. 471–488. Oxford: Oxford University Press.

Camerer, C. F. (2010). "The Case for Mindful Economics." In A. Caplin and A. Schotter (Eds.), *The Foundations of Positive and Normative Economics*, pp. 43–69. New York: Oxford University Press.

Camerer, C. F., G. Loewenstein, and D. Prelec. (2005). "Neuroeconomics: How Neuroscience Can Inform Economics." *Journal of Economic Literature* 43, 9–64.

Carey, S. (1999 [1991]). "Knowledge Acquisition: Enrichment or Conceptual Change?" In E. Margolis and S. Laurence (Eds.), *Concepts: Core Readings*, pp. 459–487. Cambridge: Bradford.

Carey, S. (2009). *The Origin of Concepts*. New York: Oxford University Press.

Carnap, R. (1938). "Logical Foundations of the Unity of Science." In O. Neurath, R. Carnap, and C. Morris (Eds.), *International Encyclopedia of Unified Science*, pp. 42–62. Chicago: University of Chicago Press.

Carnap, R. (1956a). "Empiricism, Semantics, and Ontology." In Carnap, *Meaning and Necessity* (2nd ed.), pp. 205–221. Chicago: University of Chicago Press.

Carnap, R. (1956b). *Meaning and Necessity* (2nd ed.). Chicago: University of Chicago Press.

Carroll, S. B. (2005). *Endless Forms Most Beautiful: The New Science of Evo Devo*. New York: Norton.

Cartwright, N. (1980). "The Truth Doesn't Explain Much." *American Philosophical Quarterly* 17(2), 159–163.

Cartwright, N. (1983). *How the Laws of Physics Lie*. Oxford: Clarendon.

Cartwright, N. (1999). *The Dappled World: A Study of the Boundaries of Science*. Cambridge: Cambridge University Press.

Chalmers, D. J. (2006). "Strong and Weak Emergence." In P. Davies and P. Clayton (Eds.), *The Re-Emergence of Emergence*, pp. 244–254. New York: Oxford University Press.

Chalmers, D. J. (2012). *Constructing the World*. Oxford: Oxford University Press.

Chalmers, D. J., D. Manley, and D. Wasserman (Eds.) (2009). *Metametaphysics: New Essays in the Foundations of Ontology*. Oxford: Oxford University Press.

Chomsky, N. (1959). "Review of Verbal Behavior by B.F. Skinner." *Language* 35(1), 26–58.

Chomsky, N. (2012). *The Science of Language: Interviews with James McGilvray*. New York: Cambridge University Press.

Churchland, P. (1986). *Neurophilosophy*. Cambridge, MA: MIT Press.

Churchland, P. M. (1979). *Scientific Realism and the Plasticity of Mind*. Cambridge: Cambridge University Press.

Craver, C. F. (2007). *Explaining the Brain: Mechanisms and the Mosaic Unity of Neuroscience*. New York: Oxford University Press.

Craver, C. F., and A. Alexandrova. (2008). "No Revolution Necessary: Neural Mechanisms for Economics." *Economics and Philosophy* 24, 381–406.

Craver, C. F., and L. Darden. (2013). *In Search of Mechanisms: Discoveries across the Life Sciences*. Chicago: University of Chicago Press.

Craver, C. F., and D. M. Kaplan. (2020). "Are More Details Better? On the Norms of Completeness for Mechanistic Explanation." *British Journal for the Philosophy of Science* 71(1), 287–319.

Crick, F.H. (1994). *The Astonishing Hypothesis: The Scientific Search for the Soul*, New York: Scribners.

Culp, S., and P. Kitcher. (1989). "Theory Structure and Theory Change in Contemporary Molecular Biology." *British Journal for the Philosophy of Science* 40, 459–483.

Darden, L. (1991). *Theory Change in Science: Strategies from Mendelian Genetics.* Oxford: Oxford University Press.

Darden, L., and N. Maull. (1977). "Interfield Theories." *Philosophy of Science* 44, 43–64.

Darwin, C. (1859). *On the Origin of Species* (2008 ed.). New York: Oxford University Press.

Davidson, D. (1970). "Mental Events." In L. Foster and J. Swanson (Eds.), *Experience and Theory*, pp. 79–101. London: Duckworth.

Davidson, D. (1974). "On the Very Idea of a Conceptual Scheme." *Proceedings and Addresses of the American Philosophical Association* 47, 183–198.

DeMartino, G. F. (2000). *Global Economy, Global Justice: Theoretical Objections and Policy Alternatives to Neoliberalism.* New York: Routledge.

Dennett, D. C. (1981). "Skinner Skinned." In Dennett, *Brainstorms: Philosophical Essays on Mind and Psychology*, pp. 53–70. Cambridge, MA: Bradford, MIT Press.

Dennett, D. C. (1987). *The Intentional Stance.* Cambridge, MA: Bradford, MIT Press.

Dennett, D. C. (1991). *Consciousness Explained.* Boston: Little, Brown.

Dizadji-Bahmani, F., R. Frigg, and S. Hartmann. (2010). "Who's Afraid of Nagelian Reduction?" *Erkenntnis* 73, 393–412.

Donnellan, K. (1970). "Proper Names and Identifying Descriptions." *Synthese* 21, 335–358.

Donnellan, K. (1974). "Speaking of Nothing." *Philosophical Review* 83, 3–31.

Dretske, F. (1973). "Contrastive Statements." *Philosophical Review* 82, 411–437.

Duhem, P. M. (1954). *The Aim and Structure of Physical Theory.* Princeton, NJ: Princeton University Press.

Dupré, J. (1993). *The Disorder of Things.* Cambridge, MA: Harvard University Press.

Dupré, J. (2012). *Processes of Life: Essays in the Philosophy of Biology.* New York: Oxford University Press.

Falk, R. (2009). *Genetic Analysis: A History of Genetic Thinking.* Cambridge: Cambridge University Press.

Fazekas, P. (2009). "Reconsidering the Role of Bridge Laws in Inter-Theoretic Relations." *Erkenntnis* 71, 303–322.

Feyerabend, P. K. (1993 [1975]). *Against Method* (3rd ed.). London and New York: Verso.

Firestein, S. (2012). *Ignorance: How It Drives Science.* New York: Oxford University Press.

Flanagan, O. (1991). *The Science of the Mind* (2nd ed.). Cambridge, MA: MIT Press.

Fodor, J. (1974). "Special Sciences (Or: The Disunity of Science as a Working Hypothesis)." *Synthese* 28, 97–115.

Fodor, J. A. (1999). "Let Your Brain Alone." *London Review of Books* 21.

Franklin-Hall, L. R. (2008). *From a Microbiological Point of View.* Ph.D. thesis, Columbia University, New York.

Franklin-Hall, L. R. (2016). "New Mechanistic Explanation and the Need for Explanatory Constraints." In K. Aizawa and C. Gillett (Eds.), *Scientific Composition and Metaphysical Ground: New Directions in the Philosophy of Science*, pp. 41–74. London: Palgrave MacMillan.

Franklin-Hall, L. R. (forthcoming). "The Causal Economy Account of Scientific Explanation." *Minnesota Studies in the Philosophy of Science.*

Frege, G. (1892). "On *Sinn* and *Bedeutung*." In M. Beaney (Ed.), *The Frege Reader*, pp. 251–271. Oxford: Blackwell.

Friedman, M. (1953). "The Methodology of Positive Economics." In Friedman, *Essays in Positive Economics*, pp. 3–43. Chicago and London: University of Chicago Press.

Friedman, M. (2001). *Dynamics of Reason*. Stanford, CA: CSLI Publications.

Friedman, M., and L. J. Savage. (1948). "The Utility Analysis of Choices Involving Risk." *Journal of Political Economy* 56(4), 279–304.

Garfinkel, A. (1981). *Forms of Explanation*. New Haven, CT: Yale University Press.

Giere, R. N. (1988). *Explaining Science: A Cognitive Approach*. Chicago: University of Chicago Press.

Gigerenzer, G. (2007). *Gut Feelings: The Intelligence of the Unconscious*. New York: Viking Penguin.

Gigerenzer, G. (2008). *Rationality for Mortals: How People Cope with Uncertainty*. New York: Oxford University Press.

Gillett, C. (2016). *Reduction and Emergence in Science and Philosophy*. Cambridge: Cambridge University Press.

Glennan, S. (2016). "Mechanisms and Mechanical Philosophy." In P. Humphreys (Ed.), *The Oxford Handbook of Philosophy of Science*, pp. 798–816. Oxford University Press.

Glennan, S. (2017). *The New Mechanical Philosophy*. Oxford: Oxford University Press.

Glimcher, P. W. (2011). *Foundations of Neuroeconomic Analysis*. New York: Oxford University Press.

Godfrey-Smith, P. (2003). *Theory and Reality: An Introduction to the Philosophy of Science*. Chicago: University of Chicago Press.

Godfrey-Smith, P. (2014). *Philosophy of Biology*. Princeton, NJ, and Oxford: Princeton University Press.

Goodman, N. (1955). *Fact, Fiction, and Forecast*. Cambridge, MA: Harvard University Press.

Gould, S. J. (1977). *Ontogeny and Phylogeny*. Cambridge, MA: Belknap.

Grandy, R. (1973). "Reference, Meaning, and Belief." *Journal of Philosophy* 70, 439–452.

Grene, M., and D. Depew. (2004). *The Philosophy of Biology: An Episodic History*. Cambridge: Cambridge University Press.

Griffiths, P., and K. Stotz. (2013). *Genetics and Philosophy: An Introduction*. Cambridge: Cambridge University Press.

Gul, F., and W. Pesendorfer. (2008). "The Case for Mindless Economics." In A. Caplin and A. Schotter (Eds.), *The Foundations of Positive and Normative Economics*, pp. 3–39. New York: Oxford University Press.

Hacking, I. (1983). *Representing and Intervening: Introductory Topics in the Philosophy of Natural Science*. Cambridge: Cambridge University Press.

Hall, R., and C. Hitch. (1939). "Price Theory and Business Behavior." *Oxford Economic Papers* 2, 12–45.

Hanson, N. R. (1963). *The Concept of the Positron: A Philosophical Analysis*. Cambridge: Cambridge University Press.

Harrison, G. W. (2008). "Neuroeconomics: A Critical Reconsideration." *Economics and Philosophy* 24, 303–344.

Hatfield, G. (2002). "Psychology, Philosophy, and Cognitive Science: Reflections on the History and Philosophy of Experimental Psychology." *Mind and Language* 17(3), 207–232.

Hausman, D. (1992). *The Inexact and Separate Science of Economics*. Cambridge: Cambridge University Press.

Hempel, C. G. (1945, January). "Studies in the Logic of Confirmation (I)." *Mind: A Quarterly Review of Psychology and Philosophy* 54(213), 1–26.

Hempel, C. G. (1965). *Aspects of Scientific Explanation and Other Essays in the Philosophy of Science.* New York: Free Press.

Hempel, C. G. (1966). *Philosophy of Natural Science.* Englewood Cliffs, NJ: Prentice-Hall.

Hempel, C. G., and P. Oppenheim. (1965). "On the Idea of Emergence." In C. G. Hempel (Ed.), *Aspects of Scientific Explanation and Other Essays in the Philosophy of Science,* pp. 258–264. New York: Free Press.

Hooker, C. A. (1981). "Towards a General Theory of Reduction. Part III: Cross-Categorical Reductions." *Dialogue* 20, 496–529.

Hull, D. L. (1974). *Philosophy of Biological Science.* Englewood Cliffs, NJ: Prentice-Hall.

Humphreys, P. (2016). *Emergence: A Philosophical Account.* New York: Oxford University Press.

Hutchison, T. W. (1938). *The Significance and Basic Postulates of Economic Theory.* London: Macmillan.

Hütteman, A., and A. C. Love. (2016). "Reduction." In P. Humphreys (Ed.), *The Oxford Handbook of Philosophy of Science,* pp. 460–484. New York: Oxford University Press.

Huxley, J. (1942). *Evolution: The Modern Synthesis.* London: Allen and Unwin.

Kahneman, D. (2011). *Thinking, Fast and Slow.* New York: Farrar, Straus and Giroux.

Kaplan, D. M., and C. F. Craver. (2011). "The Explanatory Force of Dynamical and Mathematical Models in Neuroscience: A Mechanistic Perspective." *Philosophy of Science* 78, 601–627.

Kim, J. (1999). *Mind in a Physical World.* Cambridge, MA: MIT Press.

Kim, J. (2005). *Physicalism, or Something Near Enough.* Princeton, NJ: Princeton University Press.

Kincaid, H. (1990). "Molecular Biology and the Unity of Science." *Philosophy of Science* 57, 575–593.

Kitcher, P. (1978). "Theories, Theorists, and Theoretical Change." *The Philosophical Review* 87(4), 519–547.

Kitcher, P. (1982). "Genes." *British Journal for the Philosophy of Science* 33, 337–359.

Kitcher, P. (1984). "1953 and All That: A Tale of Two Sciences." *The Philosophical Review* 96, 335–373.

Kitcher, P. (1985). "Darwin's Achievement." In N. Rescher (Ed.), *Reason and Rationality in Natural Science,* pp. 127–189. Lanham, MD: University Press of America.

Kitcher, P. (1993). *The Advancement of Science.* New York: Oxford University Press.

Kitcher, P. (1999). "The Hegemony of Molecular Biology." *Biology and Philosophy* 14, 195–210.

Kitcher, P., and W. C. Salmon. (1987). "Van Fraassen on Explanation." *The Journal of Philosophy* 84(6), 315–330.

Klein, C. (2009). "Reduction Without Reductionism: A Defense of Nagel on Connectability." *The Philosophical Quarterly* 59(234), 39–53.

Kordig, C. R. (1971). *The Justification of Scientific Change.* Dordrecht: Reidel.

Kripke, S. A. (1972). *Naming and Necessity.* Cambridge, MA: Harvard University Press.

Kuhn, T. S. (1962). *The Structure of Scientific Revolutions* (1st ed.). Chicago: University of Chicago Press.

Kuhn, T. S. (1977). "Objectivity, Value Judgment, and Theory Choice." In Kuhn, *The Essential Tension: Selected Studies in Scientific Tradition and Change,* pp. 320–339. Chicago: University of Chicago Press.

Kuhn, T. S. (1982). "Commensurability, Comparability, Communicability." In P. Asquith and T. Nickles (Eds.), *Proceedings of the 1982 Biennial Meeting of the Philosophy of Science Association*, pp. 669–688. East Lansing, MI: Philosophy of Science Association.

Lakatos, I. (1970). "Falsification and the Methodology of Scientific Research Programmes." In I. Lakatos and A. Musgrave (Eds.), *Criticism and the Growth of Knowledge*, pp. 91–195. Cambridge: Cambridge University Press.

Lakatos, I. (1976). *Proofs and Refutations*. Cambridge: Cambridge University Press.

Laudan, L. (1977). *Progress and Its Problems: Toward a Theory of Scientific Growth*. Berkeley and Los Angeles: University of California Press.

Leahey, T. H. (1980). *A History of Psychology*. Englewood Cliffs, NJ: Prentice- Hall.

Leahey, T. H. (2018). *A History of Psychology: From Antiquity to Modernity* (8th ed.). New York: Routledge.

Lester, R. A. (1946). "Shortcomings of Marginal Analysis for Wage-Unemployment Problems." *American Economic Review* 36, 63–82.

Lewis, D. K. (1972). "Psychophysical and Theoretical Identifications." *Australasian Journal of Philosophy* 50, 249–258.

Lewis, D. K. (1986). "Causal Explanation." In Lewis, *Philosophical Papers*, Vol. II, pp. 214–240. New York: Oxford University Press.

Lloyd, E. A. (1988). *The Structure and Confirmation of Evolutionary Theory*. Princeton, NJ: Princeton University Press.

Love, A. C., and M. J. Nathan. (2015). "The Idealization of Causation in Mechanistic Explanation." *Philosophy of Science* 82, 761–774.

Machamer, P. K. (2004). "Activities and Causation: The Metaphysics and Epistemology of Mechanisms." *International Studies in the Philosophy of Science* 18, 27–39.

Mackie, J. L. (1974). *The Cement of the Universe*. Oxford: Oxford University Press.

Maull, N. L. (1977). "Unifying Science without Reduction." *Studies in History and Philosophy of Science* 8, 143–162.

Mayr, E. (1982). *The Growth of Biological Thought*. Cambridge, MA, and London: Belknap Harvard.

Mayr, E. (1988). "The Challenge of Darwinism." In Mayr, *Toward a New Philosophy of Biology. Observations of an Evolutionist*, pp. 185–195. Cambridge, MA: Harvard University Press.

Mayr, E. (1991). *One Long Argument: Charles Darwin and the Genesis of Modern Evolutionary Thought*. Cambridge, MA: Harvard University Press.

Mayr, E. (2004). *What Makes Biology Unique?* Cambridge: Cambridge University Press.

Mayr, E., and W. B. Provine. (1980). *The Evolutionary Synthesis*. Cambridge, MA: Harvard University Press.

McCabe, K. A. (2008). "Neuroconomics and the Economic Sciences." *Economics and Philosophy* 24, 345–368.

McLaughlin, B. (1992). "The Rise and Fall of British Emergentism." In A. Beckermann, J. Kim, and H. Flohr (Eds.), *Emergence or Reduction?*, pp. 49–93. Berlin: De Gruyter.

Medawar, P. B. (1969). *The Art of the Soluble*. Harmondsworth, UK: Penguin.

Mills, S. K., and J. H. Beatty. (1979). "The Propensity Interpretation of Fitness." *Philosophy of Science* 46, 263–286.

Morange, M. (1998). *A History of Molecular Biology*. Cambridge, MA: Harvard University Press.

Morgan, M., and M. Morrison (Eds.) (1999). *Models as Mediators: Perspectives on the Natural and Social Sciences*. Cambridge: Cambridge University Press.

Morrison, M. (2012). "Emergent Physics and Micro-Ontology." *Philosophy of Science* 79, 141–166.

Nagel, E. (1961). *The Structure of Science.* New York: Harcourt Brace.

Nagel, E. (1963). "Assumptions in Economic Theory." *The American Economic Review* 53(2), 211–219.

Nagel, T. (2012). *Mind and Cosmos: Why the Materialist Neo-Darwinian Conception of Nature Is Almost Certainly False.* New York: Oxford University Press.

Nathan, M. J. (2012). "The Varieties of Molecular Explanation." *Philosophy of Science* 79(2), 233–254.

Nathan, M. J. (2014). "Causation by Concentration." *British Journal for the Philosophy of Science* 65, 191–212.

Nathan, M. J. (2015a). "A Simulacrum Account of Dispositional Properties." *Nôus* 49(2), 253–274.

Nathan, M. J. (2015b). "Unificatory Explanation." *British Journal for the Philosophy of Science* 68, 163–186.

Nathan, M. J., and G. Del Pinal. (2016). "Mapping the Mind: Bridge Laws and the Psycho-Neural Interface." *Synthese* 193(2), 637–657.

Nathan, M. J., and G. Del Pinal. (2017). "The Future of Cognitive Neuroscience? Reverse Inference in Focus." *Philosophy Compass* 12(7), 1–11.

Nicholson, D. J. (2012). "The Concept of Mechanism in Biology." *Studies in History and Philosophy of Biological and Biomedical Sciences* 43, 152–163.

Oppenheim, P., and H. Putnam. (1958). "The Unity of Science as a Working Hypothesis." In H. Feigl, M. Scriven, and G. Maxwell (Eds.), *Minnesota Studies in the Philosophy of Science*, Vol. 2, pp. 3–36. Minneapolis: Minnesota University Press.

Orzack, S. H. (2008). "Testing Adaptive Hypotheses, Optimality Models, and Adaptationism." In M. Ruse (Ed.), *The Oxford Handbook of Philosophy of Biology*, pp. 87–112. New York: Oxford University Press.

Piccinini, G., and C. F. Craver. (2011). "Integrating Psychology and Neuroscience: Functional Analyses and Mechanism Sketches." *Synthese* 183, 283–311.

Pinker, S. (1997). *How the Mind Works.* New York: Norton.

Putnam, H. (1973). "Meaning and Reference." *Journal of Philosophy* 70, 699–711.

Putnam, H. (1975a). "Philosophy and Our Mental Life." In Putnam, *Mind, Language, and Reality*, pp. 291–303. New York: Cambridge University Press.

Putnam, H. (1975b). "The Meaning of Meaning." In Putnam, *Mind, Language, and Reality*, pp. 215–271. Cambridge: Cambridge University Press.

Putnam, H. (1990). *Realism with a Human Face.* Cambridge, MA: Harvard University Press.

Quine, W. V. O. (1953). *From A Logical Point of View.* Cambridge, MA: Harvard University Press.

Quine, W. V. O. (1970). "On the Reasons for Indeterminacy of Translation." *The Journal of Philosophy* 57, 178–183.

Railton, P. (1981). "Probability, Explanation, and Information." *Synthese* 48(2), 233–256.

Robbins, L. (1932 [1984]). *An Essay on the Nature and Significance of Economic Science* (3rd ed.). London: Macmillan.

Rorty, R. (1979). *Philosophy and the Mirror of Nature.* Princeton, NJ: Princeton University Press.

Rosenberg, A. (1983). "Fitness." *Journal of Philosophy* 80, 437–447.

Rosenberg, A. (1985). *The Structure of Biological Science*. Cambridge: Cambridge University Press.

Rosenberg, A. (1992). *Economics–Mathematical Politics or Science of Diminishing Returns?* Chicago: University of Chicago Press.

Rosenberg, A. (2006). *Darwinian Reductionism: Or How to Stop Worrying and Love Molecular Biology*. Chicago: University of Chicago Press.

Russell, B. (1913). "On the Notion of Cause." *Proceedings of the Aristotelian Society* 13, 1–26.

Russell, B. (1927). *The Analysis of Matter*. London: Allen and Unwin.

Salmon, W. C. (1984). *Scientific Explanation and the Causal Structure of the World*. Princeton, NJ: Princeton University Press.

Sarkar, S. (1998). *Genetics and Reductionism*. Cambridge: Cambridge University Press.

Schaffner, K. F. (1967). "Approaches to Reduction." *Philosophy of Science* 34, 137–147.

Schaffner, K. F. (1993). *Discovery and Explanation in Biology and Medicine*. Chicago: University of Chicago Press.

Schaffner, K. F. (2006). "Reduction: The Cheshire Cat Problem and a Return to Roots." *Synthese* 151, 377–402.

Scheffler, I. (1967). *Science and Subjectivity*. Indianapolis: Bobbs-Merrill.

Sen, A. K. (1977). "Rational Fools: A Critique of the Behavioral Foundation of Economic Theory." *Philosophy and Public Affairs* 6(4), 317–344.

Shapere, D. (1966). "Meaning and Scientific Change." In R. Colodny (Ed.), *Mind and Cosmos: Essays in Contemporary Science and Philosophy*, pp. 41–85. Pittsburgh: University of Pittsburgh Press.

Shapere, D. (1974). "Scientific Theories and Their Domains." In F. Suppe (Ed.), *The Structure of Scientific Theories*. Urbana: University of Illinois Press.

Silberstein, M., and A. Chemero. (2013). "Constraints on Localization and Decomposition as Explanatory Strategies in the Biological Sciences." *Philosophy of Science* 80(5), 958–970.

Simon, H. (1963). "Problems of Methodology: Discussion." *American Economic Review: Papers and Proceedings* 53, 229–231.

Skinner, B. F. (1956). "Critique of Psychoanalytic Concepts and Theories." *Minnesota Studies in the Philosophy of Science* 1, 77–87.

Skinner, B. F. (1957). *Verbal Behavior*. New York: Appleton-Century-Crofts.

Skinner, B. F. (1974). *About Behaviorism*. (1976 ed.). New York: Vintage Books.

Skipper, R. A., and R. L. Millstein. (2005). "Thinking about Evolutionary Mechanisms: Natural Selection." *Studies in History and Philosophy of Biological and Biomedical Sciences* 36, 327–347.

Sklar, L. (1993). *Physics and Chance: Philosophical Issues in the Foundations of Statistical Mechanics*. Cambridge: Cambridge University Press.

Smith, L. D. (1986). *Behaviorism and Logical Positivism: A Reassessment of the Alliance*. Stanford, CA: Stanford University Press.

Sober, E. (1984). *The Nature of Selection: Evolutionary Theory in Philosophical Focus*. Cambridge, MA: Bradford, MIT Press.

Sober, E. (1999). "The Multiple Realizability Argument against Reductionism." *Philosophy of Science* 66, 542–564.

Sober, E. (2000). *Philosophy of Biology* (2nd ed.). Boulder, CO: Westview.

Sober, E. (2015). *Ockham's Razor: A User's Manual*. Cambridge: Cambridge University Press.

Sporns, O. (2011). *Networks of the Brain*. Cambridge, MA: MIT Press.

Sporns, O. (2012). *Discovering the Human Connectome*. Cambridge, MA: MIT Press.

Sporns, O. (2015). "Network Neuroscience." In G. Marcus and J. Freeman (Eds.), *The Future of the Brain: Essays by the World's Leading Neuroscientists*. Princeton, NJ: Princeton University Press.

Strevens, M. (2008). *Depth: An Account of Scientific Explanation*. Cambridge, MA: Harvard University Press.

Strevens, M. (2016). "Special-Science Autonomy and the Division of Labor." In M. Couch and J. Pfeifer (Eds.), *The Philosophy of Philip Kitcher*, pp. 153–181. New York: Oxford University Press.

Suppes, P. (1960). "A Comparison of the Meaning and Uses of Models in Mathematics and the Empirical Sciences." *Synthese* 12, 287–301.

Taylor, E. (2015). "An Explication of Emergence." *Philosophical Studies* 172, 653–669.

van Fraassen, B. C. (1980). *The Scientific Image*. Oxford: Clarendon.

van Fraassen, B. C. (2008). *Scientific Representation*. New York: Oxford University Press.

van Strien, M. (2014). "On the Origins and Foundations of Laplacian Determinism." *Studies in History and Philosophy of Science* 45, 24–31.

Waters, C. K. (1990). "Why the Anti-Reductionist Consensus Won't Survive the Case of Classical Mendelian Genetics." *Proceedings to the 1990 Biennial Meeting of the Philosophy of Science Association*, pp. 125–139. East Lansing, MI: Philosophy of Science Association.

Waters, C. K. (2004). "What Was Classical Genetics?" *Studies in History and Philosophy of Science* 35, 783–809.

Waters, C. K. (2006). "A Pluralist Interpretation of Gene-Centered Biology." In S. Kellert, H. Longino, and C. K. Waters (Eds.), *Scientific Pluralism*. Minnesota Studies in Philosophy of Science, Vol. XIX, pp. 190–214. Minneapolis: University of Minnesota Press.

Waters, C. K. (2007). "Causes That Make a Difference." *The Journal of Philosophy* 104(11), 551–579.

Watson, J. B. (1913). "Psychology as the Behaviorist Views It." *Psychological Review* 20, 158–177.

Weber, M. (2005). *Philosophy of Experimental Biology*. Cambridge: Cambridge University Press.

Weed, D. L. (1998). "Beyond Black Box Epidemiology." *American Journal of Public Health* 88(1), 12–14.

Weinberg, S. (1992). *Dreams of a Final Theory*. New York: Pantheon.

Weisberg, M. (2013). *Simulation and Similarity: Using Models to Understand the World*. New York: Oxford University Press.

Wilson, J. (2021). *Metaphysical Emergence*. Oxford: Oxford University Press.

Wimsatt, W. C. (2007). *Re-Engineering Philosophy for Limited Beings*. Cambridge, MA: Harvard University Press.

Wittgenstein, L. (2001). *Wittgenstein's Lectures, Cambridge 1932–1935*, A. Ambrose (Ed.), New York: Prometheus.

Wright, C., and D. Van Eck. (2018). "Ontic Explanation Is Either Ontic or Explanatory, but Not Both." *Ergo* 5(38), 997–1029.

Zador, A. (2015). "The Connectome as a DNA Sequencing Problem." In G. Marcus and J. Freeman (Eds.), *The Future of the Brain: Essays by the World's Leading Neuroscientists*, pp. 40–49. Princeton, NJ: Princeton University Press.

Index

For the benefit of digital users, indexed terms that span two pages (e.g., 52–53) may, on occasion, appear on only one of those pages.